WATER IN FOODS

First edition, Second printing

Copyright 1993 by SAMUEL A. MATZ

ISBN 0-942849-24-8

PAN-TECH INTERNATIONAL, INC.
P. O. Box 4548
McAllen, TX 78502

Printed in the United States of America

DEDICATION

This book is dedicated to
the memory of:

—L. A. Rumsey—
—T. R. Stanton—
—C. S. McWilliams—

Preface

Water is the only ubiquitous ingredient in foodstuffs. Its quantity, form, and location in these products profoundly affect their texture, appearance, and flavor. Its quality (i.e., the impurities it carries) has an important influence on the characteristics of many foodstuffs in which it has been used as an ingredient. Its chemical reactions and physical interactions with other substances present in raw materials are often the principal changes taking place during processing. These are among the reasons a reference work in the field of water science should be of interest to all food technologists.

There are several good modern texts on hydrology, hydraulics, water supply engineering, and other specialized disciplines, but the emphasis in these cases is usually too narrow to allow them to be of maximum benefit to food technologists. The purpose of *Water in Foods* is to provide a source for answers to problems which the technologist encounters in the course of his daily work. Extensive bibliographies, in which the emphasis is on recent publications (extending into 1965), will permit the reader to resolve more complex questions. With these bibliographies as guides, the food technologist can delve as deeply as he wishes into specialized aspects of the subject while, at the same time, the reader who is more interested in the broad overall picture will not be hampered by excessive detail.

The gross effects of water on food properties can be loosely divided into those related to the proportion of water substance present (the moisture content) and those caused by the impurities or additives found in ingredient water. In this book, both classes of effects will be considered. The first part of the book deals with water itself—its inherent properties, the evaluation of supplies, treatment methods, etc. The second part is concerned with the interaction of water and its impurities with other components of foods and raw materials.

Because of the current interest in saline water conversion, an up-to-date review of this subject is included, emphasizing the point of view of the food processor. There is also a discussion of waste effluent, its treatment, and its disposition since such fluids are potential sources of processing water of special interest in regions where shortages of high quality water exist or are anticipated.

SAMUEL A. MATZ

Milwaukee, Wisconsin
October 1, 1965

vii

Contents

CHAPTER 1

The Properties of Pure Water Substance

INTRODUCTION

Water is, in many respects, a unique compound. It has, and contributes to those substances which it dissolves or with which it is otherwise associated, many properties not obtainable with any other compound. Chemical reactions and physical interactions in which it participates on the molecular scale influence every gross characteristic of foodstuffs. In the present chapter, some of these special qualities of water substance which make it of special interest to the food technologist will be reviewed.

THE WATER MOLECULE

Except for a slight natural ionization which leads to the formation of minute amounts of hydrogen and hydroxyl ions, pure water consists of molecules made up of two atoms of hydrogen and one atom of oxygen. These molecules may be aggregated by weak forces into quasi-crystalline combinations whose size and form depend upon physical conditions in effect at the time.

From thermochemical data, it has been determined that the heat of formation of gaseous water from hydrogen and oxygen gases is 57.8 kcal./mole. Since the heats of dissociation of hydrogen and oxygen gases are 103.4 and 118.2 kcal./mole, respectively, the bond energy for each O–H bond is one-half of the sum $57.8 + 103.4 + 59.1$ or 110.2 kcal./mole. The chemical bonds are formed by completion of two electron pairs, each of the two unpaired electrons of the oxygen atom associating with the electron of a hydrogen atom. Four of the remaining six electrons of oxygen are much farther from the nucleus than the other two. These four plus the four electrons involved in the chemical bonds tend to form four pairs that are as far apart as they can be while still attracted to the oxygen nucleus. They form the corners of an imaginary tetrahedron.

The water molecule is schematically represented according to conventional diagramming techniques in Fig. 1. The distance of the boundary shell from the center of the atom is equal to the van der Waals' radius. The equivalent size of the molecule as determined from the density of pure ice is about 3.2 Å.

Water vapor exhibits phenomena which can be most satisfactorily explained by assuming that the molecules involved therein contain rigid, or nearly rigid, electrical dipoles, the distance between the dipoles being sig-

1

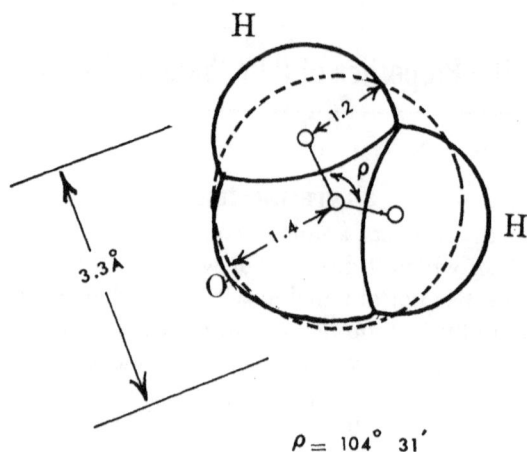

$$\rho = 104° \ 31'$$

FIG. 1. MOLECULE OF WATER VAPOR

nificantly less than the diameter of the molecule. The magnitude of these moments can be obtained from a study of the dielectric constants (Dorsey 1940). The unique properties of water with respect to molecules of similar size, such as NH_3, HF, H_2S, and CH_3OH, have been attributed (Bernal and Fowler 1933) not only to its dipole character, but also to the geometrical structure of the molecule, which is the simplest one that can form extended four-coordinated networks. The latter characteristic may in fact be more important than the dipole qualities.

According to Darling and Dennison (1940), the O—H distance in water, experimentally determined, is 0.9850 Å, the valence angle is 104° 31', and the molecule has the form of an isosceles triangle. The molecule is frequently represented as a V-shaped body in which the hydrogen nuclei are placed at the ends of two legs extending from the oxygen nucleus. The molecule possesses two positive charges centered in the hydrogen atoms and two corresponding negative charges in the region of the oxygen nucleus. It is these positive charges and their location which give rise to hydrogen bonds and account for many of the special properties of water.

In the vapor, the molecules are separated by relatively great distances and are moving at high velocity. The translational energy of these molecules is so great that the van der Waals' forces are inadequate to hold them together when they collide. Thus the vapor expands and exerts pressure in conformity with the kinetic theory of gases. In the liquid, however, the molecules are held in intimate contact with one another by the combined intermolecular forces. Each molecule in liquid water occupies a volume of 29.7 cubic angstroms which indicates a porosity between molecules of about 36.7%. The nature of the bonding forces is not well understood. Some contribution is probably made by the weak van der Waals type (with bond energies of about 5 kcal./mole) and must be, to a considerable extent, nondirectional, since the molecules are free to move in the liquid. The magnitude of the attractive forces between the molecules must be relatively great, however, since the vapor pressure of the liquid is negligible compared to the pressure which is implied by the gas laws.

HYDROGEN BONDING AND ITS CONSEQUENCES

Distribution of electrons around the hydrogen nuclei in water, as in other compounds, is asymmetrical, causing a separation of charge, or polar character (see Fig. 2). When other molecules with non-bonding outer

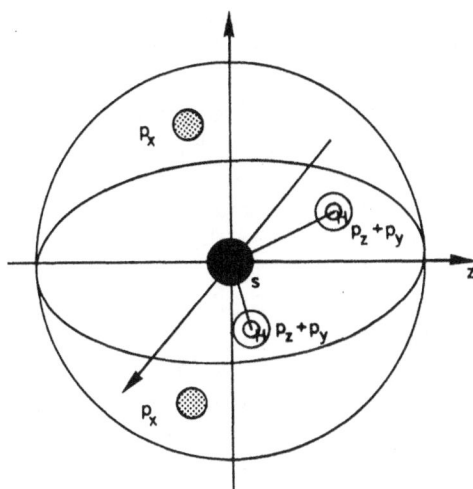

Bernal and Fowler (1933)

FIG. 2. ELECTRON DISTRIBUTION IN THE
WATER MOLECULE

electrons are present, there is a tendency for the proton to increase the symmetry of the electron distribution around it by approaching a pair of electrons which are in line with its chemical-bond electrons. In other words, the hydrogen nucleus with the two electrons constitutes a relatively weak bond. The hydrogen bond occurs commonly between hydrogen and fluorine, oxygen, nitrogen, phosphorus, and sulfur. The strength of the bond increases as the strength of attraction between the hydrogen atom and the electronegative atoms is increased and, in the series of elements just given, the strength is in decreasing order from fluorine to sulfur.

In X-ray diffraction studies of ice crystals, it was early shown that each oxygen atom is surrounded by four equidistant hydrogen atoms tetrahedrally arranged, and these in turn are bridged to other oxygen atoms. A structure of greater or lesser order is built up in this manner. The bridge is said to be symmetrical when it is essentially equidistant between the two electronegative atoms. If the proton vibrates about a position closer to one of the electronegative atoms than to the other, the bridge is asymmetrical and weaker between the hydrogen and the farther atom. The resulting attraction, when two water molecules are involved, is about six per cent as great as that of the O—H bond. Pauling (1940) lists the bond energy for $O—H \ldots O$ in water as 4.5 cal./mole. Since hydrogen-bonding is not limited to the holding together of two molecules, but may join large numbers of molecules in loose aggregates which continually break up and re-form under the influence of thermal action (Latimer and Rodebush 1920), the bonding is characterized by the fact that electrostatic forces are predominant in determining its formation and stability. As pointed out by Smith (1964), electron delocalization effects arising from the mutual polarization of the two molecules, repulsive forces between doubly-filled orbitals in the molecules and dispersion forces of the London type, must also be taken into consideration.

Several water molecules held together by hydrogen bonds, as in the solid or liquid form of the substance, form an assembly of tetrahedral groups—most molecules are bonded to four other molecules. The properties of water can be divided into two classes, depending on whether chemical bonds between the H and O atoms are broken in the action involved or whether only the hydrogen bonds are broken, leaving the H_2O molecules intact (Hendricks 1955).

By applying a statistical thermodynamic treatment based on the "flickering cluster" model of Frank and Wen (1957), some thermodynamic parameters of liquid water were derived by Némethy and Scheraga (1962). The hydrogen-bonded clusters range in average size from 91 to 25 water molecules in the range of 32° to 158°F. while the mole fraction of non-hydrogen-bonded molecules increases from 0.24 to 0.39.

The theory of hydrogen bonding is important in explaining the relationship of liquid water structure to its dielectric constant, change in viscosity with temperature and pressure, change in specific heat with temperature, change in refractive index with pressure, and the variations in a number of other less commonly encountered properties. In foodstuffs, as in many other materials, hydrogen-bonding assumes an important role in texture by loosely joining together many constituent molecules in polymer networks, forming gels, fibers, and other gross forms. Absorption and qualities of "bound" water are due to hydrogen-bonding.

STRUCTURE IN WATER AND ICE

Characteristics of water and ice which differ appreciably from those expected for liquid and solid states of molecules of approximately the same size and mass were explained at an early date as being due to an aggregation of the individual units into polymers of some sort. For example, Quincke (1905) postulated that water contained at least two species of molecules, one vastly in excess of the other. The more numerous type would be surrounded by a network of the other form. The bonding forces responsible for the stability of the network were not explained in detail by this author.

Bragg (1922) suggested a structure for ice based on the assumption that each positive ion is surrounded symmetrically by negative ions. Each oxygen atom would be at the center of gravity of four neighboring oxygen atoms from each of which it is separated by a hydrogen atom (see Fig. 3). This view provided the basis for much subsequent speculation and was supported by results of infrared, X-ray, and other experimental tech-

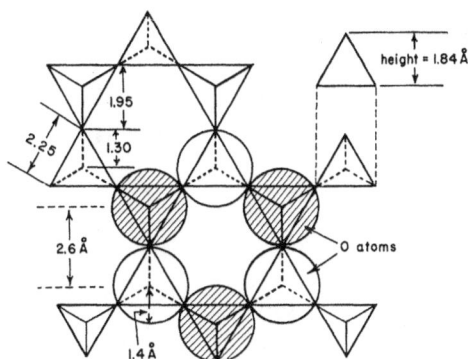

FIG. 3. TETRAHEDRAL CONFIGURATION OF ICE CRYSTAL

niques. In liquid water, the HOH units are more irregularly situated with respect to one another but they are also more densely packed, each being associated with five or more neighboring units. In this coarse and irregular network there will be present a varying number of aggregates of molecules having the crystal structure of ice. The ice structures include more vacant space than the irregularly packed form and are denser. As temperature increases from 32° to 39.2°F., the proportion of the ice-like structures decrease and the density increases. Above the higher temperature, the density of water begins to decrease because the expected and normal expansion of the liquid form is greater than the contraction that results from the change of structure.

TABLE 1

THE PHYSICAL PROPERTIES OF ORDINARY WATER SUBSTANCE[1]

Property	Value	Units	Reference
Molecular weight (on ^{12}C scale)	18.015		
Melting point, T_m	0.00	°C.	Dorsey (1940)
Triple point, T_{tr}	0.01	°C.	Dorsey (1940)
Temperature of maximum density, T_d	3.98	°C.	Dorsey (1940)
Normal boiling point, T_b	100.0	°C.	Dorsey (1940)
Critical temperature	374.15	°C.	Dorsey (1940)
Density at T_m	0.9164	gm./cm.^{3}	Lonsdale (1958)
Molar volumes:			
solid at T_m	19.66	cm.^{3}/mole	Lonsdale (1958)
liquid at T_m	18.018	cm.^{3}/mole	Dorsey (1940)
liquid at T_d	18.016	cm.^{3}/mole	Dorsey (1940)
liquid at 25°C.	18.069	cm.^{3}/mole	Dorsey (1940)
Coefficients of linear thermal expansion:			
solid at T_m	1.39×10^{-4} °C.$^{-1}$		
liquid at T_m	5.9×10^{-5} °C.$^{-1}$		
liquid at 25°C.	26.2×10^{-5} °C.$^{-1}$		
Coefficient of cubical expansion, at T_m	11×10^{-5} °C.$^{-1}$		Lonsdale (1958)
Compressibility at 25°C.	4.5×10^{-5} atm.$^{-1}$		Dorsey (1940)
Crystallographic parameters at 0°C.:			
a	4.5135	Å	Megaw (1934)
b	7.3521	Å	Megaw (1934)
c/a	1.629	Å	Megaw (1934)
length of the hydrogen bond	2.765	Å	Pimentel and McClellan (1960)
Condensation coefficient,			
water, near 0°C.	0.042		Delaney et al. (1964)
water, near 43°C.	0.027		Delaney et al. (1964)
ice between $-13°$ and $-20°$C.	0.0144		Delaney et al. (1964)
Dipole moment:			
measured in benzene solution at room temperature	1.76	Debye units	Müller (1934)
measured in the vapor at 100° to 200°C.	1.84	Debye units	Groves and Sugden (1935)
Dielectric constant at 25°C.	78.304		Malmberg (1958)
Refractive index, n_D^{20}	1.3326		Kirschenbaum (1951)
Polarizability of vapor near 100°C.	58.5	cm.^{3}/mole	Landolt and Börnstein (1959)
Viscosity at 25°C.	0.895	cp	Dirian (1962)
Surface tension at 25°C.	71.97	dynes/cm.	Dorsey (1940)

[1] Taken in part from a compilation of Némethy and Scheraga (1964).

Bernal and Fowler (1933B) put forth a very influential theory based upon the facts that water is not an ideal close-packed liquid and the molecule is asymmetrical. The following scheme was adduced: There is a form called by them "Water I" which is tridymite-like, light, and viscous; "Water II" is quartz-like, heavy, and semiviscous; "Water III" is like liquid ammonia in that it is light and nonviscous. There is no difference in the molecules themselves, but different arrangements of the same molecule exist in small regions of volume. In another paper, Bernal and Fowler (1933A) proposed an irregular four-coordinated structure to explain the large dielectric constant of water and ice and its variation with temperature and frequency. These observations were not consistent with existing theories based on molecular rotation unless most of the water molecules were not free to respond by orientation to the external electric field. According to their plan: (1) the hydrogen atoms lie on the line connecting neighboring oxygen atoms; (2) there is only one hydrogen atom on each such linkage, thus forming hydrogen bonds; and (3) each oxygen atom has two hydrogen atoms at a short distance and hence the properties of water are preserved.

It has proved difficult to get an X-ray diffraction pattern of water which could be interpreted in terms of a definite structure. Among the most thorough of the early studies was that of Morgan and Warren (1938), who obtained diffraction patterns of monochromatic Cu $K\alpha$ and Mo $K\alpha$ at 34.9°, 55.4°, 86°, 143.6°, and 181.4°F. The results were interpreted as suggesting that the tendency of a water molecule to bond itself tetrahedrally to four neighboring molecules is only partly satisfied.

More recent information makes it reasonably certain that liquid water has a modified ice structure in which the lattice is expanded anisotropically. Danford and Levy (1962) studied the X-ray diffraction of molybdenum $K\alpha$ radiation at high scattering angles from a water surface at room temperature. The nearest neighbors of the water molecules were at a distance of about 5.5% greater than in ice. A model consistent with radial distribution, neighbor distance, and other information has an ice-like framework. Each oxygen is tetrahedrally surrounded by other oxygen atoms, forming layers of puckered six-member rings. Two adjacent layers related in mirror symmetry form cavities containing interstitial water molecules. This clathrate model is said to agree with theoretical, experimental, and temperature observations.

In reviewing the current state of water structure theory, Frank (1963) stated that X-ray studies of liquid water do not support the Bernal-Fowler postulate of a quartz-like structure. However, their "uniformist" structural view has persisted in later theories, e.g. the Pople model. The latter is based on a representation of the water molecule as a rigid tetrahedral framework with two positive and two negative charge centers and the view that the hydrogen bond consists entirely of electrostatic interactions between the tetrahedra, which do not polarize or suffer distortion in the process. Since Pople permits a hydrogen bond to be bent by rotation of either the positive or negative center so that it is out of the line joining the oxygen nuclei, the model does not allow for any breaking of the bonds.

Thus the liquid is a fully bonded giant molecule at any temperature and it displays a complicated "squashed in" geometry. With this picture, plus an assumed restoring force-constant for the bending, and some rather drastic simplifying assumptions, Pople was able to calculate a radial distribution function in fair agreement with X-ray results and the temperature coefficient of the dielectric constant. According to Frank, the uniformist, or average, view underlying the Bernal-Fowler and Lennard-Jones-Pople model is still probably the most widely held, and implies that the process of averaging makes it possible to regard each water molecule at all times as doing pretty much what every other water molecule is doing.

IONIZATION

At any given time, about one molecule out of 555,000,000 molecules in a body of pure liquid water has separated into a positively charged hydrogen ion, i.e. a proton, and a hydroxyl ion, the latter being a combination of a hydrogen atom and an oxygen atom plus an additional electron. This ionization, as well as the reverse reaction, occurs spontaneously at a rate dependent upon the temperature and intensity of other sources of energy such as certain wave lengths of electromagnetic radiation, and has implications of great practical significance to the food technologist. The frequency with which specifications for raw materials and finished products include "pH" limits is testimony to this importance.

Water has an ion product of 1.0×10^{-14} at 77°F. For convenience in recording and transmitting data, the pH notation system was developed. The definition of pH is: the negative logarithm to the base 10 of the hydrogen ion concentration. As a consequence of this definition and the ion product, the logarithm of the hydroxyl ion concentration in an aqueous system is always the pH minus 14. It is generally considered more accurate to speak of hydrogen activity rather than concentration when discussing the data resulting from measurements of the apparent hydrogen ion content of any real system.

Characteristics of the two ions are quite different. The hydrogen ion is, of course, a naked proton. Since it lacks the orbiting electrons of every other ion its apparent size is exceedingly small. Its mobility of 32.5×10^{-4} cm./sec./v./cm. and its ability to penetrate between molecules is correspondingly great. Although the hydroxyl ion is quite small compared to organic molecules, it has more nearly "normal" dimensions and less capacity for catalyzing reactions. Its mobility is 17.8×10^{-4} cm./sec./v./cm. Most other ions have mobilities of around 6.7×10^{-4}.

The concept of the hydronium ion, H_3O^+ (also called the oxonium ion) was originated to explain some of the discrepancies between observations and hypotheses then current. It is doubtful, however, that such an idea really adds much to our insight into the behavior of the hydrogen ions in any situation. Experiments have shown that protons tend to flow freely through water in a way not fully measured by mobility techniques and which do not allow time for an appreciable duration of bonding between the ion and a much larger particle such as a water molecule. This is particularly true in ice which, according to Eigen and de Maeyer (1958) behaves as a protonic semiconductor. The proton mobility in ice is about 100 times that in water and is many orders of magnitude larger than that of normal ions. On the other hand, it is only about 50 times smaller than electronic mobilities in some metals and semiconductors. These authors describe proton transport in hydrogen-bonded media as being completely different from normal ionic migration. Generally, proton transport through hydrogen bonds includes two processes: (1) the formation or rearrangement of hydrogen bond structure with orientation favorable for a proton transition; and (2) the charge transfer within the hydrogen bond. The first step is rate determining in water, while the second one is decisive for the charge transport in ice. It is worthy of note that the mobility of normal ions in ice is less than one-thousandth of the value in the liquid. Eigen and de Maeyer pointed out that the reaction kinetics of protolytic systems and the fast proton transport in hydrogen-bonded systems have important implications with respect to biological problems.

BIBLIOGRAPHY

ANON. 1948. Interaction of water and porous materials. Discussions Faraday Soc. No. 3.

BADER, R. F. W., and JONES, G. A. 1963. The electron density distributions in hydride molecules. 1. The water molecule. Can J. Chem. 41, 586–606.

BASS, R. 1958. A theoretical analysis of the mechanical relaxation of single-crystalline ice. Proc. Roy. Soc. London A 247, 462–464.

BERNAL, J. D. 1958. A discussion on the physics of water and ice. Proc. Roy. Soc. London A 247, 536–538.

BERNAL, J. D., and FOWLER, R. H. 1933A. A theory of water and ionic solution, with particular reference to hydrogen and hydroxyl ions. J. Chem. Phys. 1, 515–548.

BERNAL, J. D., and FOWLER, R. H. 1933B. Pseudocrystalline structure of water. Trans. Faraday Soc. 29, 1049–1056.

BRAGG, W. H. 1922. The crystal structure of ice. Proc. Phys. Soc. London 34, 98–103.

BURTON, E. F., and OLIVER, W. F. 1936. The crystal structure of ice at low temperature. Proc. Roy. Soc. London A 153, 166–172.

BUSWELL, A. M. 1959. Fifty years of advancement in the knowledge of water substance. J. Am. Water Works Assoc. 51, 879–884.

DANFORD, M. D., and LEVY, H. A. 1962. The structure of water at room temperature. J. Am. Chem. Soc. 84, 3965–3966.

DARLING, B. T., and DENNISON, D. M.   1940.   The water vapor molecule. Phys. Rev. 57, 128–139.

DEBYE, P.   1929.   Polar Molecules.   Chemical Catalog Co., New York.

DELANEY, L. J., HOUSTON, R. W., and EAGLETON, L. C.   1964.   The rate of vaporization of water and ice.   Chem. Eng. Sci. 19, 105–114.

DORSEY, N. E.   1940.   Properties of Ordinary Water Substance.   Reinhold Publishing Corp., New York.

EIGEN, M., and DE MAEYER, L.   1958.   Self-dissociation and protonic charge transport in water and ice.   Proc. Roy. Soc. London A 247, 505–533.

ELLISON, F. O., and SHULL, H.   1955.   Molecular calculations.   I.   LCAO MO self-consistent field treatment of the ground state of H_2O.   J. Chem. Phys. 23, 2348–2350.

FISHER, J. C., HOLLOMON, J. H., and TURNBULL, D.   1949.   Rate of nucleation of solid particles in a subcooled liquid.   Science 109, 168–169.

FOX, C. S.   1952.   Water.   The Philosophical Library, New York.

FRANK, H. S.   1958.   Covalency in the hydrogen bond and the properties of water and ice.   Proc. Roy. Soc. London A 247, 481–492.

FRANK, H. S.   1963.   Questions about water structure.   Natl. Acad. Sci.   Natl. Res. Council Publ. 942, 141–155.

FRANK, H. S., and WEN, W.   1957.   Structural aspects of ion-solvent interaction in aqueous solutions: a suggested picture of water structure.   Discussions Faraday Soc. 24, 133–140.

GINZBURG, B. Z., and COHEN, D.   1964.   Calculation of internal hydrostatic pressure in gels from the distribution coefficients of non-electrolytes between gels and solutions.   Trans. Faraday Soc. 60, (part I), 185–189.

GRÄNICHER, H.   1958.   Lattice disorder and physical properties connected with the hydrogen arrangement in ice crystals.   Proc. Roy. Soc. London A 247, 453–461.

HALLETT, J., and MASON, B. J.   1958.   The influence of temperature and supersaturation on the habit of ice crystals grown from the vapour.   Proc. Roy. Soc. London A 247, 440–453.

HEEMSKERK, J.   1962.   X-ray diffraction studies on the structure of water.   Rec. Trav. Chim. 81, 904–915.

HENDRICKS, S. B.   1955.   Necessary, convenient, commonplace.   Yearbook Agr. U. S. Dept. Agr. 1955, 9–14.

HUGGINS, M. L.   1936.   Hydrogen bridges in ice and liquid water.   J. Phys. Chem. 40, 723–731.

LANGHAM, E. J., and MASON, B. J.   1958.   The heterogeneous and homogeneous nucleation of supercooled water.   Proc. Roy. Soc. London A 247, 493–504.

LATIMER, W. M., and RODEBUSH, W. H.   1920.   Polarity and ionization from the standpoint of the Lewis theory of valence.   J. Am. Chem. Soc. 42, 1419–1433.

LEWIS, G. N., and RANDALL, M.   1923.   Thermodynamics.   McGraw-Hill Book Co., New York.

LLOYD, D. J., and MORAN, T.   1934.   Pressure and the water relations of proteins.   I.   Isoelectric gelatin gels.   Proc. Roy. Soc. London A 147, 382–395.

LONSDALE, K.   1958.   The structure of ice.   Proc. Roy. Soc. London A 247, 424–434.

MEGAW, H. D. 1934. Cell dimensions of ordinary and "heavy" ice. Nature *134*, 900–901.

MORGAN, J., and WARREN, B. E. 1938. X-ray analysis of the structure of water. J. Chem. Phys. *6*, 666–673.

NAMIOT, A. Y. 1961. The double-structure of water. Zhur. Strukt. Khim. *2*, 476–477.

NÉMETHY, G., and SCHERAGA, H. A. 1962. Structure of water and hydrophilic bonding in proteins. I. A model for the thermodynamic properties of liquid water. J. Chem. Phys. *36*, 3382–3400.

PAULING, L. 1940. Nature of the Chemical Bond. Cornell University Press, Ithaca, N. Y.

PIMENTEL, G. C., and McCLELLAN, A. L. 1960. The Hydrogen Bond. W. H. Freeman, San Francisco, Calif.

PRYDE, J. A., and JONES, G. O. 1952. Properties of vitreous water. Nature *170*, 685–688.

QUINCKE, G. 1905. Ice formation and glacier grains. Ann. Physik. *18*, 1–80.

RUNDLE, R. E. 1955. The structure of ice. J. Phys. Chem. *59*, 680–682.

SIMONS, L. 1939. X-ray interference diagram of water. Soc. Sci. Fennica Commentationes Phy. Math. *10*, No. 9, 1–26.

SMITH, J. W. 1964. The hydrogen bond. Sci Progr. *52*, 97–106.

SPONSLER, O. L., BATH, J. D., and ELLIS, J. W. 1940. Water bound to gelatin as shown by molecular-structure studies. J. Phys. Chem. *44*, 996–1006.

SZENT-GYÖRGI, A. 1957. Bioenergetics. Academic Books, Ltd., London.

TAMMANN, G., and BÜCHNER, A. 1935. Supercooling capacity of water and the rate of crystallization of ice in aqueous solution. Z. anorg. u. allgem. Chem. *222*, 371–381.

VAN ECK, C. L. V. P., MENDEL, H., and FAHRENFORT, J. 1958. A tentative interpretation of the results of recent X-ray and infrared studies of liquid water and $H_2O + D_2O$ mixtures. Proc. Roy. Soc. London A *247*, 472–481.

CHAPTER 2

Sources, Composition, and Quality of
Natural Water Supplies

INTRODUCTION

Food plants are voracious consumers of water and they have, of necessity, peculiarly high standards of quality for most of the water they use. Jordan (1946) compiled a series of figures showing the normal consumption of water by factories processing many different types of foodstuffs. These data, which are reproduced in part in Table 2, indicate the need for having available a reliable and abundant supply of water. The object of this chapter is to provide a basis on which the food technologist can survey and evaluate raw water supplies.

Although several classifications of fresh water supplies are possible, the following system is thought to be particularly meaningful.

(1) Meteoritic water—Rain, snow, hail, and other forms of moisture precipitated from the atmosphere.

(2) Surface water —Any exposed body of fresh water: (a) water from rivers and other streams and from lakes and other natural reservoirs; and (b) water collected in impounding reservoirs from relatively small drainage areas.

(3) Ground water—Water existing in the saturated zone of the earth: (a) springs; (b) shallow wells; (c) deep wells; and (d) in infiltration galleries.

(4) Water from other sources: (a) demineralized brackish or sea water; and (b) re-cycled waste water.

The ultimate source of all water is, of course, meteoritic water, particularly rainfall, but direct recourse to this supply can be very costly due to the extensive area required for the collection system. The erratic nature of the supply in most areas will also necessitate large storage tanks. When it is possible to construct impounding reservoirs to collect rainfall from natural drainage formations, the resulting surface-water supply is often of high quality and relatively low in cost. Such water may not need treatment or purification of any kind before it is used in normal food processing operations. When the source is sufficiently elevated to permit distribution by gravity flow, further savings are possible in reduced operating expenses.

Few industries are so favorably situated as to be able to develop natural water supplies of this type. Most must resort to other kinds of surface water supplies or to ground waters, either directly or by purchasing water from municipal systems. Most small and many large food processing companies obtain their water from municipal distributing systems while others rely on more or less convenient sources of raw water which must be treated in the plant before it is used. In either case, an intimate knowledge of the characteristics of the raw water is desirable so that any fluctuations in quality or quantity due to seasonal or secular trends can be predicted and suitable compensating actions taken.

A desirable first step in determining the suitability of a water supply for food processing use is a sanitary survey. This is a study of environmental conditions that may affect potability of the supply. Results of the sanitary survey, in conjunction with sanitary analytical data, will furnish a reliable basis for acceptance or rejection of the supply. The Public Health Service

TABLE 2

WATER REQUIREMENTS OF FOOD MANUFACTURERS

Type of Plant	Function or Type of Food Processed	Units of Production	Water Consumption, in Gallons Per Unit
Breweries	Beer	Barrels	470
Canneries	Apricots	100 cases No. 2 cans	8,000
	Asparagus		7,000
	Beets		2,500
	Corn		2,500
	Grapefruit juice		500
	Grapefruit sections		5,600
	Green beans		3,500
	Lima beans		25,000
	Peaches		6,500
	Pears		6,500
	Peas		2,500
	Pork and beans		3,500
	Pumpkin		2,500
	Sauerkraut		300
	Spinach		16,000
	Squash		2,500
	Succotash		12,500
	Tomato products		7,000
	Tomatoes, whole		750
Meat processors	Packing plant	100 hogs killed	550
	Slaughterhouse		550
	Stockyards	Acre	160
Milk processors	Bottling works	1,000 lb. raw	250
	Cheese factory		200
	Condensing plant		150
	Dry milk factory		150
	Creamery		110
	General dairy		340
	Receiving station		180
Edible oil refiners	Soybean oil	Gallon	22
Sugar refineries	Cane sugar	Pound	0.5

[1] Adapted from Jordan (1946).

has published recommendations outlining the scope of a sanitary survey and these should be consulted before embarking on such a project. Results of an accurate survey may be more informative than an analysis and may save both delay and expense at some later time.

The sanitary analysis is the form most commonly used in the study of a public water supply and it furnishes a satisfactory foundation for judging the adequacy of the supply for food processing uses. It includes a physical, a chemical, a bacteriological, and, sometimes, a microscopical analysis. Standard methods have been published by the U. S. Public Health Service. The physical analyses include temperature, color, turbidity, odor, and taste. The specific conductance may also be reported. The chemical analyses may include: (1) acidity, alkalinity, and salinity; (2) aluminum; (3) arsenic; (4) boron; (5) calcium, magnesium, and sodium salts; (6) carbon dioxide; (7) carbonate and bicarbonate; (8) chlorides; (9) chlorine; (10) chromium; (11) colloids; (12) copper; (13) fluorides; (14) dissolved gases; (15) hardness; (16) hydrogen sulfide; (17) iodine; (18) iron; (19) lead; (20) manganese; (21) methane; (22) nitrogen, nitrate, and nitrites; (23) oxygen; (24) phenol; (25) phosphates; (26) radioactivity; (27) selenium; (28) silica; (29) silver; (30) solids; (31) sulfate; (32) detergents; (33) tin; (34) zinc; and (35) pH.

METEORITIC WATER

Meteoritic water, as it is normally collected, is quite uncontaminated and can be used for nearly all food processing applications without treatment. However, it is not, strictly speaking, "pure." There is no natural source of chemically pure water because of its remarkable solvent action.[1] Raindrops dissolve atmospheric gases as they are being formed and collect dust particles and colloidal material in their fall (as shown in Fig. 4). When the rain or melted snow comes into contact with a collecting system or a reservoir, more impurities are accumulated as a result of the solvent action. Even so, impounded meteoritic water—with impurities picked up from the collecting and storing system—is likely to be of higher quality than that obtained from any other source.

Meteoritic water which does not evaporate and which is not collected for more or less immediate use either percolates through the soil and rock strata until it merges with the body of ground water or combines into streams of increasing size or velocity. These streams and the ponds and lakes (collectively known as surface water) form the major source for municipal and food processing consumption.

---

[1] Water devoid of dissolved and suspended material is unpalatable but it would be suitable for use in food preparation.

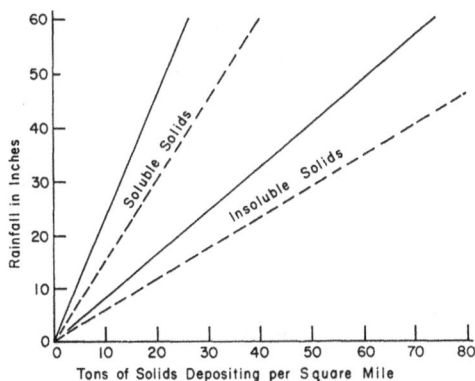

FIG. 4. SOLIDS WASHED FROM THE ATMOSPHERE IN URBAN AREAS (DASHED LINES) AND IN SUBURBAN AREAS (SOLID LINES)

SURFACE WATER

As the water precipitated from the atmosphere runs off the natural watersheds and collects in streams, ponds, lakes, swamps, etc., organic and inorganic materials are collected from the rocks and soil over which it passes, organisms grow in it, communities and individuals discharge sewage into it, and factories use it as a dump for processing wastes. The many possible avenues of contamination, in addition to the normally high levels of suspended matter and the elevated indices of color and turbidity, make it necessary to treat almost all surface waters before use. Mountain streams or lakes arising in uninhabited areas at high altitudes may contain water which is sufficiently pure to use without extensive treatment, but these sources are usually not accessible to food manufacturers.

Natural purification of surface water occurs as the result of sedimentation, oxidation, radiation (i.e., sunlight), dilution, ingestion of contaminating substances or organisms by higher animals, and chemical reactions.

Sedimentation proceeds more rapidly in relatively quiescent bodies such as ponds than it does in flowing streams. As a result, visible contaminants are maintained for a longer period of time in streams. However, bacterial pollution may be neutralized more quickly in running water because of the increased reaeration provided by the mixing of water and air. Rapidly flowing streams tend to spread pollutants for considerable distances downstream while the area of contamination remains more restricted (though it is more concentrated) in lakes. A high velocity contaminated stream may pollute the water for a distance of several miles from the point of its entry into a lake. In lakes, particularly, the spread of pollution varies with the seasons, being much affected by winds, thermal currents, and the like in summer, while during the winter in temperate and frigid zones the number of bacteria becomes more uniform throughout, as a result of the stabilizing effect of the ice covering.

The temperature of surface water is subject to wide fluctuations. The maximum summer temperature may be from 36° to as much as 60°F. above the minimum winter temperature. These fluctuations must be taken into account when calculating the energy requirements for heating or cooling process water.

The mineral content of surface water is extremely variable depending to a large extent on geographical and geological factors. The expected range of some of the mineral constituents is shown in Table 3. The quality of

TABLE 3

DISSOLVED MINERAL CONSTITUENTS IN A GROUP OF 98 RIVERS[1]

Component	Concentration[2]	
	Minimum	Maximum
Hardness (Ca and Mg)	15	589
Calcium (Ca)	11	408
Magnesium (Mg)	3	181
Sodium plus Potassium (Na + K)	4	774
Bicarbonate (HCO_3^-)	14	256
Chloride (Cl)	1	702
Sulfate (SO_4^-)	4	473
Nitrate (NO_3^-)	0.1	10
Iron (Fe)	0.02	3
Silica (SiO_3)	8	48

[1] Adapted from Camp (1963).
[2] As parts per million $CaCO_3$ except iron (as parts per million iron) and silica (as parts per million SiO_3).

the finished product can be deleteriously, or, in a few cases, advantageously, affected by excess hardness or by the presence of a significant amount of certain "non-hardness" ions. The seasonal range as well as the average content of mineral substances should be known before a source of water is selected. If a source having excessive hardness must be used, plans must be made to install a softening unit in the treatment plant.

The amount of water which can be withdrawn from a stream can be estimated by reference to gagings of its flow over a period of years. The Water Supply Papers of the United States Geological Survey contain a vast amount of detailed information on stream flow.

GROUND WATER

Ground water has been defined as that existing in the saturated zone of the earth, thus excluding water which may be found intermittently in higher levels of the soil. Substantially all ground water originates as meteoritic water which penetrates the surface of the ground, percolates through connecting voids in various soil strata or in porous rocks, and flows along the surface of (or between) the more impenetrable layers of rock. The contribution to ground water of sewage and other aqueous wastes deposited on or in the earth is relatively small. Surface water can make a very substantial contribution to the ground water supply but the total supply is generally rather closely tied to the amount of rainfall in the originating area, which may be far removed from the point of extraction.

In limestone regions, large underground streams and pools may be found, but these are seldom available for commercial use, so that springs, wells, and infiltration galleries put down in gravel or sand deposits or in porous rock formations are the common means of tapping ground water supplies.

In the terminology of ground-water hydrology, voids are understood to be the spaces between loose or cemented particles forming the different strata of soil and rock. In the saturated zone, these voids are filled with water, while at higher levels they may be filled with either gas or water. Collectively, voids constitute the porosity, which can be expressed either as the percentage of the total volume occupied by voids or as the ratio of voids to total volume. Available ground-water storage capacity is directly proportional to porosity but the usefulness of the storage capacity depends upon the ability of the soil to let water flow to a point where it can be extracted—the so-called permeability—and on specific yield. The latter term is defined as the percentage of total volume of water drained under the pull of gravity. The specific yield is always less than the porosity, as some water will be retained in the formation by surface forces (Suter 1964).

The porosity of gravel and sand deposits will vary from nearly 25 to 40% or more. The porosity of rock strata range from almost zero in some granites to perhaps as much as 30% in very open-grain sandstone.

Wells may be dug, bored, driven, or drilled. The principal differences in waters are between those from deep and shallow wells. Many deep water supplies are of sufficiently high quality to be usable either for drinking or for most food plant uses without treatment, while shallow wells more rarely yield water of such purity. Most well and spring waters are free of turbidity. Microorganisms are absent or present at very low levels in water from deep wells, especially if the aquifer is protected by overlying impervious strata, but shallow wells, infiltration galleries, and some springs may yield appreciable numbers of bacteria. When these are present, they sometimes originate from seepage of surface water down the side of the shaft, but unobstructed passages for contaminated water may also exist as continuous crevices in fragmented or cavernous rock extending to the surface, especially in limestone areas.

Deep well waters are usually colorless and odorless except for the occasional appearance of hydrogen sulfide in wells located near oil fields or in shale. They are often high in dissolved carbon dioxide and relatively high in dissolved minerals. The temperature of these waters is quite constant. Wells from 30 to 60 ft. deep have temperatures about 2° to 3°F. above the mean annual air temperature, or between 40° and 50°F. throughout most parts of the United States. The temperature increases about 1°F. for each additional 64 ft. of depth.

In some locations, infiltration galleries provide an efficient means for collecting a water supply. They consist of a series of horizontal parallel conduits placed in a porous soil formation and draining into a receiving well. The conduits or pipes are porous or perforated and are placed in trenches filled in with gravel. Favorable formations for the location of infiltration galleries are sand and gravel strata situated below a stable water table. They are usually placed in close proximity (but not nearer than 50 ft.) to a lake or stream, along the bank or under the bed, but they can also be used to collect ground water when a dependable water table is present close to the surface. Purification of the water occurs as a result of its filtration through the sand between source and collection point. Water from an infiltration gallery will have about the same quality as water from a shallow well dug in the same area. The ground above the infiltration gallery must be protected from pollution by animals or sewage (Salvato 1958).

WATER OBTAINED FROM MUNICIPAL SYSTEMS

The cost of pumping water is generally much less than the price of water delivered from municipal systems, so that large users often find it more economical to develop their own water supplies. Cost of municipal water averaged throughout the United States is 12.3 cents per 1,000 gal. at the consumer's tap (Hoak 1962). The average cost of supply from river water is about five cents. Some industries rely on dual supplies to make up the deficiencies of their own source, purchasing water for sanitary uses and fire-fighting purposes while using their own source for untreated water intended for non-sanitary consumption. In these dual systems, cross connections must be avoided since check valves and similar devices cannot be relied upon to prevent contamination of the sanitary supply, even if the valves are used in series. When city water is mixed with the plant's private supply in making up shortages, a direct connection can be prevented by dropping the municipal water into a make-up tank from a point high enough to render backsiphonage impossible at all water levels. The local health authorities usually have rules governing hookups of private water systems with the municipal supply.

Most municipal systems use surface water supplies and nearly all treat the water, as shown in Table 4, which contains data obtained in a 1960 survey of the U. S. Public Health Service.   Due to public health regula-

TABLE 4

ANALYSIS OF SUPPLY SERVED TO COMMUNITIES OF 25,000 OR MORE[1]

	Communities	
Supply	Number	Percentage of the Total
Source		
Surface	355	56.5
Ground	144	22.9
Surface and ground	129	20.6
Condition		
Treated	598	95.2
Untreated	6	1.0
Treated and untreated	24	3.8
Ownership		
Public	515	85.3
Private	86	14.2
Public and private	3	0.5
Not stated	24	

[1] Jenkins and Owen (1963).

TABLE 5

TRACE ELEMENTS IN FINISHED WATER SUPPLIES[1]

Element	Observed Concentrations, p.p.m.			1962 USPHS DWS Limit
	Low	High	Mean	
Antimony (Sb)	0.001	0.1	0.25	None
Arsenic (As)	None	0.1	0.01	0.01[2], 0.05[3]
Barium (Ba)	0.0007	0.9	0.049	1.0[2]
Beryllium (Be)	0.00001	0.0007	0.00013	None
Boron (B)	None	1.0	0.1	None
Cadmium (Cd)	0.0004	0.06	0.0082	0.01[3]
Chromium (Cr^{6+})	0.0003	0.04	0.0032	0.05
Cobalt (Co)	0.0003	0.03	0.0056	None
Copper (Cu)	None	0.6	0.032	1.0[2]
Fluoride (F)	None	10	0.73	1.4–2.4[4]
Lead (Pb)	0.001	0.4	0.017	0.05[3]
Molybdenum (Mo)	0.0004	9.2	0.0096	None
Nickel (Ni)	0.0004	0.4	0.0117	None
Selenium (Se)	None	0.01	0.008	0.01[3]
Silver (Ag)	None	2 ppb	0.13 ppb	50 ppb
Strontium-90 (Sr^{90})	0.1[5]	3.3[5]	1.0[5]	10[5]
Tin (Sn)	0.0003	0.03	0.006	None
Vanadium (V)	0.0004	0.07	0.006	None
Zinc (Zn)	0.06	7.0	1.33	5.0[2]

[1] U. S. Public Health Service data quoted by Taylor (1963).
[2] These values should not be exceeded if other more suitable supplies are available.
[3] Constitutes grounds for rejection of supply.
[4] See U. S. Public Health Service Standards for the significance of temperature variations.
[5] Strontium-90 values in $\mu\mu c/l$.

tions in effect throughout the country, all municipal systems can be relied upon to supply safe drinking water, but potability is not the sole criterion of suitability for food processing uses since the water may be excessively hard (see Fig. 5) turbid, or colored, or have off-odors or -flavors, which interfere with product quality when it is incorporated into a food. Trace elements may be very prominent in these effects. Table 5 indicates the range of these components found to exist in finished water supplies.

IMPURITIES IN NATURAL WATERS AND THEIR SIGNIFICANCE

Foreign constituents occurring in water supplies may affect the turbidity, color, odor, taste, corrosiveness, hardness, and safety. All of these characteristics are important to the food processor. All natural water supplies contain dissolved mineral matter but amounts and relative proportions of the different ions may vary greatly. Bicarbonates, sulfate, and chloride are the most common anions, while calcium, magnesium, and sodium are found more frequently and in greater quantity than the many other cations which may be present. Dissolved gases originating in the atmosphere are always present. Of these, nitrogen is of little importance and is practically never determined in water analysis because it is inert in food processing procedures. Less carbon dioxide is found in surface than in ground water.

Hydrogen sulfide is a most offensive contaminant in natural waters. Not only is the odor pervasive and undesirable, but the compound also makes the water more corrosive to iron, steel, copper, and brass. It can originate from either bacterial action or a chemical source. Its odor is apparent in the cold at 0.5 p.p.m. and becomes very obnoxious at a concentration of 1 p.p.m. About 70 p.p.m. of the gas is a physiological irritant and it is considered highly poisonous above 700 p.p.m.

Methane is a result of microbiological activity. It is found in relatively few ground waters. Under rare conditions, this gas may actually constitute a fire and explosion hazard, especially in partially filled tanks or at other points where it can collect in enclosed spaces.

Colors of natural waters range from very light yellow to dark brown. Color is usually due mainly to organic substances, but not all such compounds contribute to color. Black and Christman (1963) showed that the materials responsible for color in water exist primarily in colloidal suspension. Colloidal or organically-bound iron and manganese enhance color and are often, but not always, found in relatively high concentrations in strongly colored natural waters. Iron and manganese compounds are found in most soils, gravel, shale, and sandstone. The essentially insoluble oxides are taken up by water in which the oxygen tension has been re-

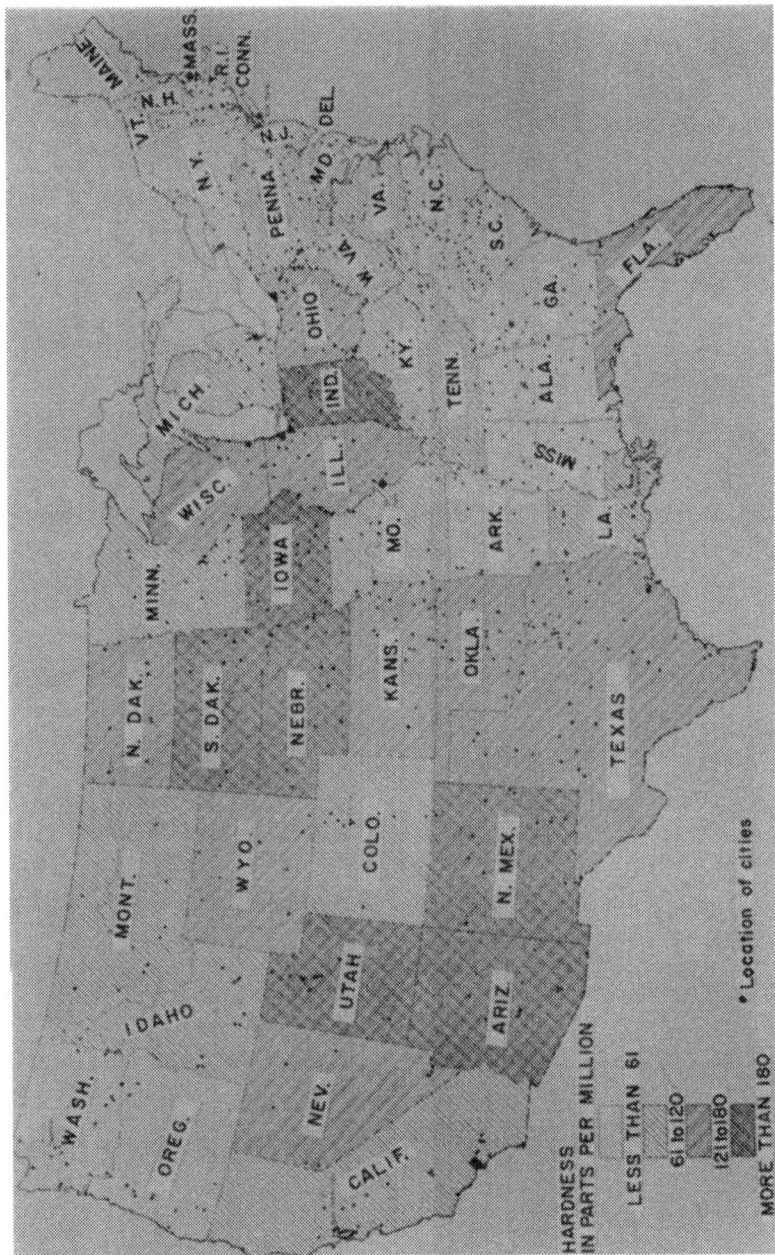

Fig. 5. Map Showing Weighted Average Hardness, by States, of Finished Water for Public Supplies for 1,315 of the Larger Cities in the United States During 1952.

duced by decomposing organic matter or by other processes, or by water containing substantial amounts of carbonic acid. Ferrous iron is found as the bicarbonate in alkaline well or spring waters while the sulfate predominates in acidic drainage. In the presence of air and in the absence of interfering substances, the bicarbonate will precipitate out a floc of ferric hydroxide.

Iron in excess of about 0.3 to 0.5 p.p.m. will cause water to appear rusty. Although dissolved iron in even larger concentrations is not physiologically harmful, it can be very objectionable for many food processing applications and renders the water unacceptable as a beverage ingredient. It will also stain equipment, particularly porcelainized receptacles. Red water occurs as a result of corrosion and is due to the presence of minute flecks of ferric compound. Manganese accompanies iron in most cases and has rather similar effects.

Mineral acids can be present in sufficient amounts in waters of low pH to impart a distinctly sour taste. Bitter or astringent tastes can be due either to ionic constituents or to organic contaminants.

Undissolved materials in water may include inorganic substances such as clay, rock flour, silt, calcium carbonate, silica, ferric hydroxide, and sulfur, or organic contaminants such as finely divided animal or vegetable matter, oils, fats, greases, and microorganisms. Size of particles may be in the colloidal range or as large as grains of sand. The larger granules drop out rapidly in quiescent water and are seen as sediment while the more finely divided particles are responsible for the turbidity. Microorganisms and higher organisms may be brought in from the supply, especially if it is surface water, or may grow in the reservoirs. Definite effects on color, odor, turbidity, and taste result.

Perhaps as many as 186 genera of organisms (Fair and Whipple 1927) are found in water, but only a small number of these are of economic importance. The latter include about ten genera of green algae, blue-green algae, diatoms, and protozoa.

When odors described as aromatic or geranium-like are observed, diatoms or, perhaps, protozoa are to blame. These organisms can cause the development of fishy odors. Grassy odors may also be due to algae. The number and metabolic activity of the organisms have a considerable influence on the character of the odor. According to Gainey and Lord (1952), *Asterionella* and *Tabellaria* cause a noticeable "earthy" odor when they are present in small numbers, but the predominant note has more of an aromatic character when 500 to 1,000 standard units per milliliter are present. At this level and below, only a few unusually sensitive people will find the odor objectionable. In the range of about 1,000 to several thousand, a definite geranium-like smell is observed which gradually changes to a fishy odor at even higher levels. Above 2,000 standard units per milliliter, many persons will regard the odor as being quite unpleasant. It is obviously desirable to entirely prevent or eliminate such organisms, if this is practical. Some, like *Synura*, are particularly objectionable even in small numbers, and contribute a bitter after-taste to the water subsequent to chlorination. This taste is difficult to eliminate by ordinary processing techniques.

BIBLIOGRAPHY

ACKERMAN, E. A., and LÖF, G. O. G. 1959. Technology in American Water Development. Resources for the Future, Inc., Washington, D. C.

ALDRICH, D. G., JR. 1962. Water for the growing state of California. Food Technol. *16*, No. 7, 24–27.

ANON. 1949. Hydrology Handbook. American Society of Civil Engineers, New York.

ANON. 1952. Water Quality Criteria. State Water Pollution Control Board, Sacramento, Calif.

ANON. 1953. Industrial water use. J. Am. Water Works Assoc. *45*, 289–300.

ANON. 1955. Water. Yearbook Agr. U. S. Dept. Agr. *1955*.

ANON. 1956. Manual on Industrial Water. Am. Soc. for Testing Materials Special Tech. Publ. *148B*.

ANON. 1962. Tentative method for carbon chloroform extract (CCE) in water. J. Am. Water Works Assoc. *54*, 223–227.

BABBIT, H. E., DOLAND, J. J., and CLEASBY, J. L. 1962. Water Supply Engineering. 6th Edition. McGraw-Hill Book Co., New York.

BLACK, A. P., and CHRISTMAN, R. F. 1963. Characteristics of colored surface waters. J. Am. Water Works Assoc. 55, 753–770.

CAMP, T. R. 1963. Water and Its Impurities. Reinhold Publishing Co., New York.

CHILDS, E. C. 1964. The movement of soil water. Endeavour *23*, 81–84.

FAIR, G. M., and GEYER, J. C. 1954. Water Supply and Waste Water Disposal. John Wiley and Sons, New York.

FAIR, G. M., and WHIPPLE, M. C. 1927. Whipple's Microscopy of Drinking Water. John Wiley and Sons, New York.

FLINN, A. D., WESTON, R. S., and BOGERT, C. L. 1927. Waterworks Handbook. 3rd Edition. McGraw-Hill Book Co., New York.

GAINEY, P. L., and LORD, T. H. 1952. Microbiology of Water and Sewage. Prentice-Hall, Englewood Cliffs, N. J.

HARDENBERGH, W. A. 1945. Water Supply and Purification. International Textbook Co., Scranton, Penna.

HICKS, T. G. 1957. Pump Selection and Application. McGraw-Hill Book Co., New York.

HICKS, T. G. 1958. Pump Operation and Maintenance. McGraw-Hill Book Co., New York.

HIRSHLEIFER, J., DE HAVEN, J. C., and MILLIMAN, J. W. 1960. Water Supply: Economics, Technology, and Policy. Univ. of Chicago Press, Chicago.

HOAK, R. D. 1962. Are we really running out of water? Trusts and Estates 5, No. 4, 35–40.

HYNES, H. B. N. 1960. The Biology of Polluted Waters. Liverpool Univ. Press, Liverpool, England.

JENKINS, K. H., and OWEN, H. W. 1963. The 1960 USPHS summary of municipal water facilities in communities of 25,000 or more. J. Am. Water Works Assoc. 55, 35–41.

JOHNSTONE, D. and CROSS, W. P. 1949. Elements of Applied Hydrology. Ronald Press, New York.

JORDAN, H. E. 1946. Industrial requirements for water. J. Am. Water Works Assoc. 38, 65–68.

LAW, F. 1961. Some useful ways of analyzing upland catchment data. J. Institution Water Engineers 15, 159–167.

LINCK, C. J. 1964. Planned ground water exploration. J. Am. Water Works Assoc. 56, 418–422.

LINSLEY, R. K., JR. and BRANZINI, J. B. 1964. Water-Resources Planning. Scranton Publishing Co., Chicago.

LINSLEY, R. K., JR., KOHLER, M. A., and PAULHUS, J. L. H. 1949. Applied Hydrology. McGraw-Hill Book Co., New York.

MANSFIELD, W. W. 1956. The use of hexadecanol for reservoir evaporation control. Southwest Research Institute, San Antonio, Texas.

MILLER, D. W., GERAGHTY, J. J., and COLLINS, R. S. 1962. Water Atlas of the United States. Water Information Center, Port Washington, New York.

PAULSEN, C. G. 1950. Quality of the surface waters of the United States— 1946. U. S. Geol. Surv. Water Supply Paper 1050.

PICTON, W. L. 1960. Water use in the United States 1900–1980. Water and Sewerage Industry and Utilities Division, U. S. Business and Defense Services Administration, Department of Commerce.

RAINWATER, F. H., and THATCHER, L. L. 1953. Methods for the collection and analysis of water samples. U. S. Geol. Surv. Water Supply Paper No. 1454.

SALVATO, J. A. JR. 1958. Environmental Sanitation. John Wiley and Sons, New York.

SAVILLE, T. 1917. The nature of color in water. J. New England Water Works Assoc. 31, 78–80.

SHAPIRO, J. 1963. Natural coloring substances of water and their relation to inorganic components. Am. Chem. Soc. Div. Water Waste Chem. Preprints 1963, (March–April) 6–12.

SMITH, S. C. and CASTLE, E. N. 1964. Economics and Public Policy in Water Resource Development. Iowa State Univ. Press, Ames, Iowa.

STEEL, E. W. 1953. Water Supply and Sewerage. McGraw-Hill Book Co., New York.

SUTER, M. 1964. Effect of voids and permeability on ground-water flow condition. Water Works and Wastes Engineering 1, 41–45.

SVERDRUP, H. R., JOHNSON, M. W., and FLEMING, R. H. 1942. The Oceans— Their Physics, Chemistry, and General Biology. Prentice-Hall, Englewood Cliffs, N. J.

TAYLOR, F. B. 1963. Significance of trace elements in public finished water supplies. J. Am. Water Works Assoc. 55, 619–623.

THRESH, J. C., and BEALER, J. F. 1925. The Examination of Waters and Water Supplies. P. Blakiston's Sons, Philadelphia.

TIMBLIN, L. O., JR., MORAN, W. T., and GARSTKA, W. U. 1957. Use of molecular layers for reservoir evaporation reduction. J. Am. Water Works Assoc. *49*, 841–850.

WELSH, G. B., and THOMAS, J. F. 1960. Significance of chemical limits in USPHS drinking water standards. J. Am. Water Works Assoc. *52*, 289–300.

WISLER, C. O. and BRATER, E. F. 1959. Hydrology. John Wiley and Sons, New York.

CHAPTER 3

Tests of Water Quality

INTRODUCTION

The literature of the technology of water analysis is quite voluminous, showing an understandable concern of chemists with the safety and quality of our water supplies. Over the past few decades standardized techniques for determining nearly all of the important water quality factors have emerged and gained general acceptance. Some of these have attained official, or quasi-official, status through inclusion in *Standard Methods for the Examination of Water and Wastewater* (Anon. 1960B), which contains nearly all of the tests referred to in the *Drinking Water Standards* (Anon. 1962A). It is customary to divide tests for water quality into the categories of physical, bacteriological, chemical, and radioactivity determinations, the classification which will be followed in the subsequent discussion.

CHEMICAL TESTS

According to the *Drinking Water Standards,* the analytical methods used to determine compliance are those specified in *Standard Methods for the Examination of Water and Wastewater* and the following: (1) barium—according to Rainwater and Thatcher (1953); (2) carbon chloroform extract (CCE)—according to Middleton *et al.* (1959); (3) radioactivity—(Anon. 1959A and 1959B); and (4) selenium—according to Magin *et al.* (1960).

Some of the analyses which have been standardized are: acidity, alkalinity, aluminum, arsenic, boron, calcium, carbon dioxide, chloride, chlorine (residual chlorine), chlorine demand, chromium, color, specific conductance, copper cyanide, fluoride, grease, hardness, iodide, iron, lead, lignin, magnesium, manganese, albuminoid nitrogen, ammonia nitrogen, nitrate nitrogen, nitrite nitrogen, oxygen, oxygen consumed, pH value, phenol, phosphate, potassium, residue, selenium, silica, sodium, sulfate, sulfide, sulfite, tannin and lignin, taste and odor, temperature, turbidity, and zinc. There are alternate standard or tentative methods for some of these factors. Since the standard methods are readily available, there is no point in reproducing them here.

Samples of fresh water for analysis are customarily measured by volume at the prevailing temperature and not by weight. As a practical matter, corrections for temperature and specific gravity are almost never made except for samples of boiler salines, brines, and other high-solids liquids. The analyst will assume that one liter weighs one kilogram, that one U. S. gallon weighs 8.33 lb., and that one British imperial gallon weighs 10 lb. (Nordell 1961).

There are only four basic units used for reporting the results of water analyses, although there are six different names for them. These are: (1) parts per hundred thousand (parts/100,000); (2) parts per million (p.p.m.), occasionally expressed as milligrams per liter or grams per cubic meter; (3) grains per U. S. gallon (g.p.g.). One grain is 1.43×10^{-4} lb. (0.06480 gm.) and one U. S. gallon of water weighs 8.33 lb.; and (4) grains per imperial gallon (g.p.g. imp.), the number of grains of a substance per one British imperial gallon of sample (an imperial gallon of water weighs 10 lb.).

The following conversion factors will be found to be useful: (1) To convert grains per gallon to parts per million, multiply by 17.1; (2) to convert parts per million to grains per gallon, multiply by 0.0583; (3) to convert grains per imperial gallon to parts per million, multiply by 14.3; and (4) to convert parts per million to grains per imperial gallon, multiply by 0.07.

Among the most frequently performed chemical tests are those which determine the mineral content and the residual chlorine. Seven methods for determining residual chlorine are in common use by water analysts:

(1) **Orthotolidine method.**—This colorimetric determination is probably the most widely used of the residual chlorine tests. Nitrites, ferric compounds, manganic compounds, organic iron compounds, lignocellulose, and microorganisms can interfere with the test.

(2) **Amperometric method.**

(3) **Iodometric method.**—Another colorimetric method which can be made more precise than the orthotolidine technique, especially when the residual is greater than one part per million.

(4) **Flash test.**—A colorimetric test in which the color must be standardized. It is not applicable in the presence of oxidized manganese.

(5) **Enzymatic method.**—This involves the measurement of the time necessary for milk containing the unknown amount of chlorine to clot when it is incubated with the proteolytic enzyme papain. The clotting time under uniform conditions is a function of the concentration of residual chlorine.

(6) **Drop dilution test.**—A modification of the orthotolidine test for use in the field. It is especially applicable when the chlorine concentrations are greater than 10 parts per million as reached in water main sterilization.

(7) **Continuous recording of residual chlorine.**—It depends on depolarization of a copper electrode immersed in an electrolytic cell through which the chlorinated water flows continuously. The electric current passing through the cell is proportional to the concentration of residual chlorine in water. Factors affecting the sensitivity of the test include electrode deterioration, temperature, pH, and chloramines. The concentrations of dissolved solids normally found in potable water are said to have no significant effect on the functioning of the cell (Babbitt *et al.* 1962).

Lishka *et al.* (1963) reported the results of a collaborative study by 69 laboratories intended to evaluate current analytical methods for the determination of minerals in water. Each participant received a sample made by adding carefully measured amounts of calcium sulfate, potassium nitrate, magnesium chloride, sodium nitrite, and sodium bicarbonate to distilled, de-ionized water. Data obtained in this survey, which was begun in 1961, were compared with the results of similar studies made in 1956 and 1958. Based on the results of the three surveys, the authors concluded that the following methods were the most acceptable: calcium, EDTA titration; magnesium, calculated; chloride, mercuric nitrate titration; sulfate, turbidimetric; hardness, EDTA titration; nitrite, diazotization; nitrate, phenoldisulfonic (inadequate, further research desirable); sodium, flame photometer; potassium, flame photometer; and alkalinity, electrometric endpoint.

Simplified procedures using pre-measured reagents, special indicators, and electronic instrumentation have been developed as alternatives to the standard techniques (Anon. 1963).

As mentioned by Lishka *et al.* (1963), the standard phenoldisulfonic test for nitrate appears to be unsatisfactory. Jenkins and Medsker (1964) described a new brucine-based colorimetric method for the determination of nitrate in both fresh and saline water. It is said to give highly reproducible results in the range of 0.05 to 0.8 mg./l. nitrate nitrogen. In this range the color produced bears an essentially linear relationship to the nitrate concentration. The tentative brucine procedure listed in the Standard Methods is not recommended for nitrate concentrations of less than one milligram per liter. The Jenkins and Medsker technique makes use of controlled heating and chloride masking. Unlike the phenoldisulfonic determination, the brucine test is not much subject to chloride interference,, at least within the range of 0 to 20 gm./l. of chloride.

Lombardi (1964) reported on the use of di-β-naphthylthiocarbazone (Dinaphthizone) compared with dithizone as an analytical reagent for the determination of trace metals in natural waters. This author concluded that Dinaphthizone was the reagent of choice for determining trace metals in alkaline lake brines, by reason of the superior extraction of the complexed metals and the unreacted reagent and also because no pretreatment is needed except dilution and filtering. In the extraction of copper and silver from alkaline samples, the reagent gives much better extractions than does dithizone. The latter is superior, however, for samples having a pH below 8.

A modification of the Standard Methods EDTA titrimetric method for calcium and magnesium permitting the simultaneous determination of these elements was developed by Katz and Navone (1964). Magnesium ions in the sample are precipitated as the hydroxide by the addition of NaOH until a pH of 12 to 13 is reached. Calcium ions in the unfiltered sample are complexed by titration with EDTA using Eriochrome Blue S. E. [disodium-1-hydroxy-4-chlorobenzene (2 azo-2-1,8 dihydroxynaphthalene-3,6-disulfonic acid)] as the indicator. After titration of the calcium ions, the solution is acidified to dissolve the magnesium hydroxide completely. The solution is buffered, a few drops of Eriochrome Black T indicator are added, and the magnesium is titrated with EDTA. The addition of the second indicator improves the magnesium endpoint.

According to Johnson et al. (1964), the standard method is suitable for measuring sulfite but the evolution method gives incomplete recoveries for sulfide. These authors tried a new method based on separation of zinc sulfide by inverse filtration followed by titration in an inert atmosphere. It was capable of recovering 0.2 mg. of sulfide from liter samples of sea water with an accuracy of about 0.01 mg. The inverse filtration method may be used to determine sulfide in the range of 0.2 to 53 mg./l. with a relative standard deviation averaging about 1.4%. Analysis of an equilibrated sample requires only 15 to 30 min. Although designed for use with estuarine and sea waters, the method is also suitable for the determination of sulfide in water, waste water, and possibly sewage.

The standard method for fluoride determination in potable water specifies a distillation step to remove interference prior to spectrographic analysis. Kelso et al. (1964) found that adsorption of fluoride by an anionic resin (e.g. Dowex 2X-8, 50–100 mesh, chloride form) followed by stripping with a solution containing beryllium ion eliminates the need for distillation. The accuracy and precision obtained with the new method are comparable to those obtained by standard distillation procedures but the saving in time and equipment is considerable.

BIOLOGICAL OXYGEN DEMAND

Non-living organic matter in water tends to become the substrate for bacterial growth and reproduction. The metabolic processes of these microorganisms utilize the oxygen dissolved in the water and give off carbon dioxide. The biological oxygen demand, B.O.D., or the total amount of oxygen which is taken up, is thus an indication of the extent to which the water is polluted with organic substances derived from living matter, i.e., sewage and the like. Of course, B.O.D. determinations do not give any information on the type or kind of organic compounds which are present.

The B.O.D. has been defined by the American Public Health Association as "the oxygen in parts per million required during the stabilization of the organic matter by aerobic bacterial action." It is also stated that "Complete stabilization requires more than 100 days at 68°F., but such long periods of incubation are impractical in any but research investigations, consequently a much shorter period of incubation is used. Incubation for 1, 2, 5, 10, or 20 days at 68°F. is customary, and the 5-day period is recommended as the standard procedure. Conversion of the data from one incubation period to another, or from one temperature to another, may then be made."

The bacterial consumption of oxygen may be partially offset by the photosynthetic processes of microscopic plant life (such as algae and diatoms) which produce oxygen and consume carbon dioxide.

According to McKee and Wolf (1963), B.O.D. is important from a pollution standpoint only so far as it decreases dissolved oxygen, or produces septicity or subsequent growth of saprophytic bacteria which increases the turbidity or other undesirable characteristics of streams. In a slow, sluggish stream, a five-day B.O.D. of five milligrams per liter might be sufficient to produce anaerobic conditions, while a swift mountain stream might easily handle 50 mg./l. without appreciable reduction of the concentration of dissolved oxygen. Excessive B.O.D. can result in suffocation of fish.

BACTERIOLOGICAL ANALYSES

There are three more or less traditional bacteriological tests applied to water: (1) counts of growth on gelatin plates incubated at 68°F.; (2) counts on agar plates incubated at 99°F. for 24 hr.; and (3) the coliform count. The latter criterion has been abandoned as a standard in favor of the "Most Probable Number (MPN) of E. coli" which has been defined as "that bacterial density, which, if it had actually been present in the sample under observation, would more frequently than any other have given the observed analytical results."

Plate counts at 68° or 99°F. are not particularly important as indicators of water safety but they are useful as routine quality control tests in the various water treatment procedures and as means for determining the sanitary conditions of basins, filters, etc. The coliform test is the ultimate measure of bacteriological safety of water supplies. Tests can be of three

successive types: presumptive (indicating the presence or absence of coliform bacteria and/or other organisms producing gas); positive (indicating the presence or absence of coliform bacteria only); and confirmed
(confirmed presence of coliform bacteria). The first of these tests is more
often applied to surface waters while the last two represent more defini-
tive tests applicable to water for drinking, food processing, swimming, etc.
The presumptive test requires incubation at 99°F. and examination after
48 hr. of serial dilutions of the sample inoculated into a standard lactose
broth. Gas production is a positive result. As the name indicates, the test
is not specific for *E. coli*. Results supplement those obtained from a sanitary survey in establishing hygienic quality of the water sample.

The coliform group includes "all aerobic and facultative anaerobic
Gramnegative, nonspore-forming bacilli which ferment lactose with gas
formation." Bacteria having these characteristics include organisms that
differ not only in their biochemical and serological characteristics but also
in their natural sources and habitats and, therefore, in their public health
significance. The organism in this group which serves as the prime indicator of fecal pollution is *Escherichia coli* since it is characteristically a
major part of the flora of normal human and animal digestive tracts.
Subgroups such as intermediate-aerogenes-cloacae (I.A.C.), which also
ferment lactose with the production of gas, may originate from fecal discharges but they are more frequently found on certain types of vegetation
and in soil. Another subgroup includes some plant pathogens and other
organisms of obscure habitat.

PLANKTON

It is sometimes desirable to have an estimate of the quantity of plank-
tonic organisms in a reservoir or other water source. The number of these
organisms is ordinarily too small for direct microscopic counting, although
visual enumeration of the amounts present in counting chambers can be
undertaken after a preliminary concentration by centrifuging or filtering.
The concentration of plankton is usually reported in areal standard units
or volumetric standard units. The areal standard units are determined by
measuring the number of 400-micron-square areas covered by the cross-
sectional areas of the microorganisms from a given volume of water. The
volumetric standard unit is equal to 8,000 cubic microns. Although the
latter is doubtless more accurate, it involves a difficult third-dimensional
measurement, and is seldom used.

A standard method of concentration consists of filtering a given volume
of water through a specific grade of sand, followed by a re-suspending of
the collected organisms in a smaller volume of water. A Sedgwick-Rafter
counting chamber with a depth of one millimeter and an area of 1,000 sq.

mm. (one milliliter capacity) is filled with the concentrated sample. The ocular of the microscope is fitted with a Whipple net micrometer. With the aid of a stage micrometer, the ocular is adjusted so that the smallest squares of the net micrometer equal 400 square microns, an areal standard unit. This procedure greatly facilitates the measuring of the area of the plankton organisms in the counting chamber. Since the microorganisms of a given species are fairly constant in size, it is necessary only to measure the area of a few, calculate the average, and transpose this value into areal standard units (Gainey and Lord 1952).

The number of microorganisms in a plankton can also be determined directly by drawing a given volume of the water through a filter of appropriate type (Anon. 1963A) and counting, with the aid of a microscope, those organisms trapped by the membrane.

PHYSICAL EXAMINATION

Turbidity, color, and odor are routinely determined when a complete water analysis is conducted. These tests do not directly measure the safety of water but they are related to consumer acceptance of water and of the foods and beverages made from it. Public Health Standards indicate that water becomes objectionable to a considerable number of consumers when limits of five units of turbidity, 15 units of color, and threshhold odor number of three are exceeded.

Turbidity can be measured by using a Jackson candle turbidimeter (see Fig 6). The water sample is poured into a calibrated tube until a depth is reached at which the image of the flame is completely obscured. Low turbidity samples cannot be measured in this way because the height of

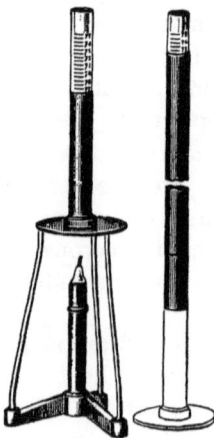

FIG. 6. JACKSON TURBIDIMETER

the column of water which would be necessary to obscure the flame image becomes unrealistic. However, their turbidity can be estimated by comparing the sample with standards or by using photoelectric instruments. Dilutions of highly turbid samples can be made to bring them within range of the turbidimeter readings.

Turbidity can be continuously measured and automatically recorded by instruments such as the one shown in Fig. 7. The illustrated sensing unit

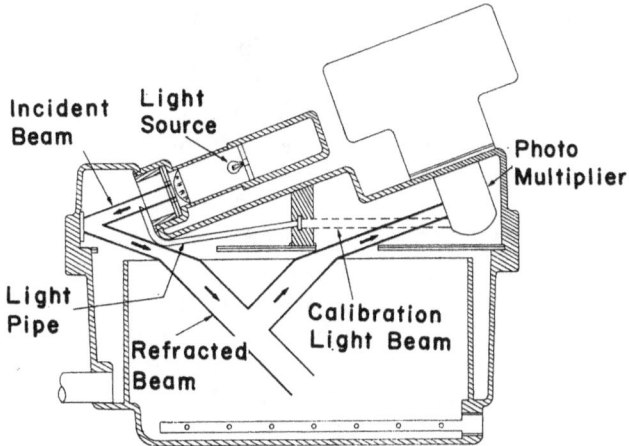

Courtesy of the Foxboro Co.

FIG. 7.   THE FOXBORO TURBIDIMETER

is operated in conjunction with a power supply and a recorder or indicator. It acts by measuring Tyndall light, i.e., light reflected at $90° \pm 10°$ from the surface of particles suspended in a liquid. The quantity of light reflected (detected by a photomultiplier tube) is a direct function of the concentration of particles in suspension and, therefore, a direct measure of turbidity. Water or other fluid flows continuously through the sample chamber at a rate of one liter per second. Fresh fluid rises through the liquid in the reservoir and spills over a weir into a drain sump without turbulence at the surface to present a still liquid surface to both the incident and reflected beams of light. This apparatus is continuously adjustable from a low of 0.5 turbidity units full-scale to a high of 1,000. Since the turbidity curve reverses itself between about 400 and 500 turbidity units, turbidity can be read only between 0 to 400 and 500 to 1,000 T.U.

Turbidity is a measure of the extent to which light passing through the water is reduced by suspended material and can be considered to be the inverse of "clarity." It can be due to the presence of microorganisms, organic debris, silica, clay, silt, or industrial wastes of various kinds. For many food uses, a rather high level of turbidity can be tolerated, but in preparation of carbonated beverages and the like, water of high clarity must be used.

Color of water is defined as that difference in hue caused only by those substances which are actually in solution. It can be due to mineral or vegetable pigments of natural origin or it can be caused by soluble organic or inorganic materials from sewage or industrial effluents. To obtain a quantitative estimation of "color," the analyst visually compares the sample with platinum-cobalt standard solutions which are supposed to approximate the natural hue of water. Semipermanent standards of colored glass tubes or discs are also used. The standard numerical value is the color unit or the equivalent of the color produced by one milligram of platinum per liter of solution. Doubtless photoelectric colorimeters will ultimately replace visual comparison tests.

RADIOACTIVITY DETERMINATIONS

Limits of three micromicrocuries per liter of radium-226 and 10 $\mu\mu$c of strontium-90 per liter have been established. These levels considerably exceed those observed for water from any normal source. Generally speaking, analyses for radioactivity of water used in food processing or as an ingredient would not be required. Manuals of procedures for radioactive determinations are listed in the bibliography of this chapter (Anon. 1959A and 1959B).

BIBLIOGRAPHY

ANON. 1949. American Society for Testing Materials Standards on Industrial Water. American Society for Testing Materials, Philadelphia.

ANON. 1952. Water Quality Criteria. Calif. Water Pollution Control Board Publ. 3.

ANON. 1954. Manual on Industrial Water. American Society for Testing Materials Special Tech. Bull. 148A.

ANON. 1955A. Standard Methods for the Examination of Water, Sewage, and Industrial Wastes. Tenth Edition. American Public Health Association, New York.

ANON. 1955B. Water. Yearbook Agr. U. S. Dept. Agr. 1955.

ANON. 1959A. Laboratory Manual of Methodology, Radionuclide Analysis of Environmental Samples. U. S. Public Health Serv. Tech. Rept. R59-6.

ANON. 1959B. Methods of Radiochemical Analysis. World Health Organization Tech. Rept. 173.

ANON. 1960A. Official Methods of Analysis. Ninth Edition. Association of Official Agricultural Chemists. Washington, D. C.

ANON. 1960B. Standard Methods for the Examination of Water and Wastewater. 11th Edition. American Water Works Assoc., New York.

ANON. 1962A. Drinking Water Standards. U. S. Public Health Serv. Publ. 956.

ANON. 1962B. Recommended Procedures for the Bacteriological Examination of Sea Water and Shellfish. 3rd Edition. American Public Health Assoc., New York.

ANON 1962C. Tentative method for carbon chloroform extract (CCE) in water. J. Am. Water Works Assoc. 54, 223–227.

ANON. 1963A. Microbiological Analysis of Water and Milk—Application Data Manual ADM-10. Millipore Filter Corp., Bedford, Mass.

ANON. 1963B. Water Analysis Procedures. Hach Chemical Co., Ames, Iowa.

BABBITT, H. E., DOLAND, J. J., and CLEASBY, J. L. 1962. Water Supply Engineering. 6th Edition. McGraw-Hill Book Co., New York.

BAYLIS, J. R. 1935. Elimination of Taste and Odor in Water. McGraw-Hill Book Co., New York.

BLACK, A. P., and HANNAH, S. A. 1963. Measurement of low turbities in water. Am. Chem. Soc. Div. Water Waste Chem. Preprints 1963, 1–5.

BROWN, H. B. 1957. The meaning, significance, and expression of commonly measured water quality criteria and potential pollutants. Louisiana State Univ. Eng. Expt. Sta. Bull. 58.

CAMP, T. R. 1963. Water and Its Impurities. Reinhold Publishing Corp., New York.

COHEN, J. M., KAMPHAKE, L. J., HARRIS, E. K., and WOODWARD, R. L. 1960. Taste threshold concentrations of metals in drinking water. J. Am. Water Works Assoc. 52, 660–670.

EVANS, A. F. 1961. Taste and odor control for palatable water. J. New Engl. Water Works Assoc. 75, 77–80.

FAIR, G. M., and WHIPPLE, M. C. 1927. Whipple's Microscopy of Drinking Water. John Wiley and Sons, New York.

GAINEY, P. L. 1950. An Introduction to the Microbiology of Water and Sewage for Engineering Students. Burgess Publishing Co., Minneapolis, Minn.

GAINEY, P. L., and LORD, T. H. 1952. Microbiology of Water and Sewage. Prentice-Hall, Englewood Cliffs, N. J.

JENKINS, D., and MEDSKER, L. L. 1964. Brucine method for determination of nitrate in ocean, estuarine, and fresh waters. Anal. Chem. 36, 610–612.

JENKINS, K. H., and OWEN, H. W. 1963. The 1960 USPHS summary of municipal water facilities in communities of 25,000 or more. J. Am. Water Works Assoc. 55, 35–41.

JOHNSON, C. R., McCLELLAND, P. H., and BOSTER, R. L. 1964. Rapid volumetric determination of sulfide in estuarine and sea water. Anal. Chem. 36, 301–302.

KATZ, H., and NAVONE, R. 1964. Method for simultaneous determination of calcium and magnesium. J. Am. Water Works Assoc. 56, 121–123.

KELSO, F. M., MATTHEWS, J. M., and KRAMER, H. P. 1964. Ion-exchange method for determination of fluoride in potable waters. Anal. Chem. 36, 577–579.

KRAMER, H. P., and KRONER, R. C. 1959. Cooperative studies in laboratory methodology. J. Am. Water Works Assoc. 51, 607–620.

KRAMIG, G. 1964. Use of the total carbon analyzer for pollution control. Part 1. The Analyzer 5, No. 4, 11–14.

KRONER, R. C., BALLINGER, D. G., and KRAMER, H. P. 1960. Evaluation of laboratory methods for analysis of heavy metals in water. J. Am. Water Works Assoc. 52, 1466–1470.

LISHKA, R. J., KELSO, F. S., and KRAMER, H. P. 1963. Evaluation of methods for determination of minerals in water. J. Am. Water Works Assoc. 55, 647–656.

LOMBARDI, O. W. 1964. Di-β-naphthylthiocarbazone (Dinaphthizone) compared with dithizone as an analytical reagent for the determination of trace metals in natural waters. Anal. Chem. 36, 415–418.

MAGIN, G. B., THATCHER, L. L., RETTIG, S., and LEVINE, H. J. 1960. Suggested modified method for colorimetric determination of selenium in natural waters. J. Am. Water Works Assoc. 52, 1199–1205.

McKEE, J. E., and WOLF, H. W. 1963. Water Quality Criteria. 2nd Edition. State Water Quality Control Board, Sacramento, Calif.

MIDDLETON, F. M., ROSEN, A. A., and BURTTSCHELL, R. H. 1959. Manual for Recovery and Identification of Organic Chemicals in Water. Robert A. Taft Sanitary Engineering Center, Cincinnati, Ohio.

NORDELL, E. 1961. Water Treatment for Industrial and other Uses. 2nd Edition. Reinhold Publishing Corp., New York.

PRESCOTT, S. C., WINSLOW, C. E. A., and McCRADY, M. H. 1946. Water Bacteriology. John Wiley and Sons, New York.

RAINWATER, F. H., and THATCHER, L. L. 1953. Methods for the collection and analysis of water samples. U. S. Geological Survey Water Supply Paper No. 1454.

RUBIN, H., and SWARBRICK, R. H. 1961. Rapid fluorine analysis by wideline nuclear magnetic resonance. Anal. Chem. 33, 217–220.

SCHAFFER, R. B. 1964. Use of the total carbon analyzer for pollution control. Part III. The Analyzer 5, No. 4, 14–15.

SIGWORTH, A. A. 1964. Consistency of results with threshold odor tests. J. Am. Water Works Assoc. 56, 297–300.

SMIT, J. 1948. Microbiology of drinking water and sewage. Ann. Rev. Microbiol. 11, 435–452.

SUCKLING, E. V. 1944. The Examination of Water and Water Supplies. P. Blakiston's Son and Co., Philadelphia.

TAYLOR, E. W. and BURMAN, N. P. 1964. The application of membrane filtration techniques to the bacteriological examination of water. J. Applied Bact. 27, 294–303.

THEROUX, F. R., ELDRIDGE, E. F., and MALLMAN, W. L. 1943. Laboratory Manual for Chemical and Bacterial Analysis of Water and Sewage. McGraw-Hill Book Co., New York.

THRESH, J. C., and BEALE, J. F. 1925. The Examination of Water and Water Supplies. P. Blakiston's Son and Co., Philadelphia.

WEIL, B. H. 1948. Bibliography on water and sewage analysis. Georgia St. Eng. Expt. Sta. Spec. Rept. No. 28.

WEIL, B. H., MURRAY, P. E., REID, G. W., and INGOLS, R. S. 1948. Bibliography of Water and Sewage Analysis. Georgia Inst. Technol. State Expt. Sta. Spec. Rept. 28.

WUHRMANN, K. 1964. River bacteriology and the role of bacteria in self-purification of rivers. In Principles and Applications in Aquatic Microbiology. Edited by H. Heukelekian and M. C. Dondero. John Wiley and Sons, New York.

CHAPTER 4

Public Health Service Drinking
Water Standards

BACKGROUND AND GENERAL CONDITIONS OF THE STANDARDS

In 1962, the U. S. Department of Health, Education, and Welfare published a new version of Drinking Water Standards. These standards were prepared by the Advisory Committee on Revision of the Public Health Service 1946 Drinking Water Standards and were promulgated for use in administration of interstate quarantine regulations. Although they are ostensibly intended to apply only to water used on common carriers engaged in interstate commerce, they have, in practice, a very much wider influence. The American Water Works Association regards these standards as minimum requirements for protection and well-being of individuals and communities, and works for establishment of the standards as criteria of quality for all public water supplies in the United States. They should also be regarded as criteria for the quality of water added to foods during their manufacture or processing.

In drawing up the latest revision of the standards, the Committee observed the following guidelines:

(1) The proposed standards should be discussed widely and due cognizance should be given to international and other standards of water quality before a final report is submitted; (2) a new section on radioactivity should be added; (3) greater attention should be given to chemical substances being encountered increasingly in both variety and quantity in water sources; (4) in establishing limits for toxic substances, intake from food and air should be considered; (5) the rationale employed in determining the various limits should be included in an appendix; (6) the proposed format, with the exceptions noted above, should not differ greatly from the present standards; (7) the standards should be generally acceptable and should be applicable to all public water supplies in the United States, as well as those supplies used by carriers subject to the Public Health Service regulations; (8) the following two types of limits used in previous editions should be continued: (a) Limits which, if exceeded, shall be grounds for rejection of the supply. Substances in this category may have adverse effects on health when present in concentrations above the limit. (b) Limits which should not be exceeded whenever more suitable supplies are, or can be made. available at reasonable cost. Substances

in this category, when present in concentrations above the limit, are either objectionable to an appreciable number of people or exceed the levels required by good water quality control practices; and (9) these limits should apply to water at the free-flowing outlet of the ultimate consumer.

In view of the importance of the standards to food manufacturers, the requirements will be discussed in detail in the following paragraphs.

The standards specify that the water supply shall be obtained from the most desirable source which is feasible, and effort should be made to prevent or control pollution of the source. If the source is not adequately protected by natural means, the supply shall be adequately protected by treatment. Adequate protection by natural means is defined as involving one or more of the following processes of nature that produces water consistently meeting the requirements of the standards: dilution, storage, sedimentation, sunlight, aeration, and the associated physical and biological processes which tend to accomplish natural purification in surface waters and, in the case of ground waters, the natural purification of water by infiltration through soil and percolation through underlying material and storage below the ground water table. Adequate protection by treatment is defined as meaning any one or any combination of the controlled processes of coagulation, sedimentation, absorption, filtration, disinfection, or other processes which will produce a water consistently meeting the requirements of the standards. This protection also includes processes which are appropriate to the source of supply; works which are of adequate capacity to meet maximum demands without creating health hazards, and which are located, designed, and constructed to eliminate or prevent pollution; and conscientious operation by well-trained and competent personnel whose qualifications are commensurate with the position and acceptable to the reporting agency (the official State health agency or its designated representative) and the certifying authority (the Surgeon General of the U. S. Public Health Service or his duly authorized representative).

The standards direct that frequent sanitary surveys be made of the water supply system in order to locate and identify health hazards which might exist in the system. The manner and frequency of making these surveys, and the rate at which the health hazards are to be removed, shall be in accordance with a program approved by the reporting agency and the certifying authority. According to definition, the water supply system includes the works and auxiliaries for collection, treatment, storage, and distribution of the water from the sources of supply to the free-flowing outlet of the ultimate consumer.

Approval of water supplies is to be dependent in part upon: (1) enforcement of rules and regulations to prevent development of health hazards; (2) adequate protection of water quality throughout all parts of the system, as demonstrated by frequent surveys; (3) proper operation of the water supply system under responsible charge of personnel whose qualifications are acceptable to the reporting agency and the certifying authority; (4) adequate capacity to meet peak demands without development of low pressures or other health hazards; and (5) records of laboratory examinations showing consistent compliance with water quality requirements of these standards.

Responsibility for conditions existing in a water supply system is held by (1) the water purveyor from the source of supply to the connection to the customer's service piping, and (2) the owner of the property served and the municipal, county, or other authority having legal jurisdiction from the point of connection in the customer's service piping to the free-flowing outlet of the ultimate consumer. As can be seen from the preceding requirements, the food manufacturer would be considered responsible for the condition and quality of the water in his service piping before and after any treatment applied to it within the plant.

PHYSICAL AND CHEMICAL CHARACTERISTICS

The appropriate reporting agency and the certifying authority will specify the frequency and manner of sampling. However, it is recommended by the Public Health Service that, under normal circumstances, samples should be collected one or more times per week from representative points in the distribution system and examined for turbidity, color, threshold odor, and taste. Drinking water should contain no impurity which would cause offense to the senses of sight, taste, or smell. Turbidity should not exceed 5 units, color should not exceed 15 units, and threshold odor number should be 3 or less for general use water.

Chemical analyses for the constituents listed in Tables 4 and 5 are usually performed semiannually unless the responsible authority indicates that a particular analysis may be omitted because the substance in question is expected from good evidence to be absent or below levels of concern. Where a presumption of unfitness exists because of the presence of undesirable substances in previous analyses or for other reasons, more frequent examinations should be made and there should be an exhaustive sanitary survey to determine the source of contamination. The standards state that available and acceptable source water analyses performed in accordance with standard methods may be used as evidence of compliance if the concentration of a substance is not expected to increase in processing and distribution.

Table 7 includes limiting concentrations for substances which may adversely affect health. The presence of any of these substances in concen-

TABLE 6

DESIRABLE LIMITS FOR CHEMICAL SUBSTANCES IN POTABLE WATER[1]

Substance	Maximum Concentration,[2] Mg./L.
Alkyl benzene sulfonate (ABS)	0.5
Arsenic (As)	0.01
Chloride (Cl)	250
Copper (Cu)	1
Carbon chloroform extract (CCE)	0.2
Cyanide (CN)	0.01
Iron (Fe)	0.3
Manganese (Mn)	0.05
Nitrate[3] (NO_3^-)	45
Phenols	0.001
Sulfate ($SO_4^=$)	250
Total dissolved solids	500
Zinc (Zn)	5

[1] Anon. 1962.
[2] These concentrations should not be exceeded if, in the judgment of the reporting agency and Certifying Authority, other more suitable supplies are or can be made available.
[3] In areas in which the nitrate content of water is known to be in excess of the listed concentration, the public should be warned of the potential dangers of using the water for infant feeding.

TABLE 7

ESSENTIAL LIMITS FOR CHEMICAL SUBSTANCES IN WATER SUPPLIES[1]

Substance	Maximum Concentration,[2] Mg./L.
Arsenic (As)	0.05
Barium (Ba)	1.0
Cadmium (Cd)	0.01
Chromium, hexavalent (Cr^{+6})	0.05
Cyanide (CN^-)	0.2
Lead (Pb)	0.05
Selenium (Se)	0.01
Silver (Ag)	0.05

[1] Anon. 1962.
[2] Presence of the substances in excess of the maximum concentration constitutes grounds for rejection of the supply.

TABLE 8

LIMITS ON FLUORIDE CONCENTRATION IN WATER SUPPLIES[1]

Annual Average of Maximum Daily Air Temperatures[2]	Recommended Control Limits for Artificially Fluoridated Water			Maximum Limit, Mg./L.
	Lower, Mg./L.	Optimum, Mg./L.	Upper, Mg./L.	
50.0–53.7	0.9	1.2	1.7	2.4
53.8–58.3	0.8	1.1	1.5	2.2
58.4–63.8	0.8	1.0	1.3	2.0
63.9–70.6	0.7	0.9	1.2	1.8
70.7–79.2	0.7	0.8	1.0	1.6
79.3–90.5	0.6	0.7	0.8	1.4

[1] Anon. 1962.
[2] Based on temperature data obtained for a minimum of five years.
[3] Average concentrations in excess of these limits constitute grounds for rejection of the supply.

trations greater than those listed is grounds for rejection of the supply. Table 6 lists limiting concentrations for substances which can cause the supply to be objectionable. The presence of these substances in excess of the concentrations listed is justification for seeking a supply of better quality. When such a supply is not available health authorities should be consulted regarding the suitability of using the substandard supply.

Table 8 describes the limits for fluoride content of water. Where fluoride addition is practiced, the average fluoride concentration should be kept within the upper and lower control limits. If only natural fluoride is present, the concentration should not average more than the appropriate upper limit in the table. Average concentrations of fluoride in excess of those shown in the right hand column constitute grounds for rejection of the supply.

Radioactivity

In conformity with the view that exposure of humans to ionizing radiation should be kept at the lowest practicable level, the standards contain limits for radioactive materials in drinking water. Limits for total exposure from all sources, including water, are based on the recommendation of the Federal Radiation Council (September 13, 1961) that "Routine control of useful applications of radiation and atomic energy should be such that expected average exposures of suitable samples of an exposed population group will not exceed the upper value of Range II (20 $\mu\mu c$/day of Radium-226 and 200 $\mu\mu c$/day of Strontium-90)." When the concentrations in the water of the latter substances do not exceed 3 $\mu\mu c$/l. and 10 $\mu\mu c$/l., respectively, the Standards indicate that the supply is to be considered acceptable without surveillance of other sources of radioactivity caused by Ra^{226} and Sr^{90}. Water supplies containing Ra^{226} and Sr^{90} in larger amounts can be approved if the total ingestion of radioactivity from all sources does not exceed the limits recommended by the Federal Radiation Council.

If Strontium-90 and alpha-emitters (e.g. Radium-226) are present at negligibly small fractions of the specific limits, the water supply is acceptable even through gross beta concentrations are as high as 1,000 $\mu\mu c$/l. Gross beta concentrations in excess of this amount are grounds for rejection of the supply except when more complete analyses indicate that concentrations of nuclides are not likely to cause exposures greater than the limits established by the Federal Radiation Council.

ANALYTICAL PROCEDURES

Most of the analytical tests used in policing the requirements of the Public Health standards are those given in the latest edition of "Standard

Methods for the Examination of Water, Sewage, and Industrial Wastes" (Anon. 1955). The analysis for barium is that of Rainwater and Thatcher (1953). The test for carbon chloroform extract is conducted according to the procedure of Middleton *et al.* (1962). In determining selenium, the method of Magin *et al.* (1960) is used.

Tests for radioactive materials follow the procedures contained in the "Laboratory Manual of Methodology, Radionuclide Analysis of Environmental Samples" (1959) and "Methods of Radiochemical Analysis" (1959B).

BIBLIOGRAPHY

ANON. 1955. Standard Methods for the Examination of Water, Sewage, and Industrial Wastes. American Public Health Association, New York.

ANON. 1959A. Laboratory Manual of Methodology, Radionuclide Analysis of Environmental Samples. Robert A. Taft Sanitary Engineering Center, Public Health Service Tech. Rept. *R59-6.*

ANON. 1959B. Methods of Radiochemical Analysis. World Health Organization Tech. Rept. *173.*

ANON. 1962. Public Health Service Drinking Water Standards. U. S. Department of Health, Education, and Welfare, Washington, D. C.

MAGIN, G. B., THATCHER, L. L., RETTIG, S., and LEVINE, H. J. 1960. Suggested modified method for colorimetric determination of selenium in natural water. J. Am. Water Works Assoc. 52, 1199–1205.

MIDDLETON, F. M., ROSEN, A. A., and BURTTSCHELL, R. H. 1959. Manual for Recovery and Identification of Organic Chemicals in Water. Robert A. Taft Sanitary Engineering Center, Cincinnati, Ohio.

MIDDLETON, F. M., ROSEN, A. A., and BURTTSCHELL, R. H. 1962. Tentative method for carbon chloroform extracts (CCE) in water. J. Am. Water Works Assoc. 54, 223–227.

RAINWATER, F. H., and THATCHER, L. L. 1953. Methods for the collection and analysis of water samples. U. S. Geological Survey Water Supply, Paper *1454.*

CHAPTER 5

Treatments Applied to Water

INTRODUCTION

The treatment needed by a water supply depends not only upon its original composition and quality but also upon the use for which the finished water is intended. Nearly all natural water supplies must be treated in some manner before they can be used either in food or as drinking water. Water from a municipal system which is perfectly acceptable for drinking and suitable for most food processing applications may have to be further treated, as by ion-exchange techniques, before it will be adequate for certain specialized food and beverage uses. Sedimentation and chlorination are usually the minimum treatments necessary for water from the best sources. Coagulation and filtration are required more often than not, and softening or other advanced treatments are becoming more prevalent in municipal systems. Types and amounts of dissolved and suspended organic and inorganic impurities as well as the microbiological flora determine the kind and extent of treatment. Table 9 summarizes the recommendations of the U. S. Public Health Service on the relationship of bacterial quality to minimum treatment.

The purposes for which treatments are applied may be conveniently divided into three categories: (1) treatments intended to alleviate organoleptic deficiencies such as color, turbidity, off-tastes, and odors; (2) treatments designed to eliminate health hazards such as toxic chemicals and disease-causing microorganisms and viruses; and (3) treatments needed for making the water more satisfactory for specialized uses.

It is not within the scope of this book to discuss in detail the equipment used for treating water in municipal plants. However, some food plants have their own treatment facilities for producing water of satisfactory quality from raw water supplies or for improving municipal water which has one or more undesirable characteristics. In this chapter, principles of water treatment which might be applied by food plants and equipment available for these applications will be discussed.

The first step in purifying water from a contaminated source is a rough screening operation to remove gross articles such as tin cans, fish, weeds, etc. Sets of parallel bars, screens, and the like are placed at the entrance to intake cribs. From the inlet, pumps raise the water to the level of treating tanks and filters. Subsequent flow is generally gravity powered.

TABLE 9

A CLASSIFICATION OF WATERS BY CONCENTRATION OF COLIFORM BACTERIA AND
TREATMENT REQUIRED TO RENDER THE WATER OF SAFE SANITARY QUALITY[1]

Group Number	Maximum Permissible Average MPN Coliform Bacteria per Month	Treatment Required
1	Not more than 10% of all 10 ml. portions positive; approximately less than 2.2 coliform bacteria per 100 ml.	No treatment required of underground water, but a minimum of chlorination required of surface water.
2	Not more than 50/100 ml.	Simple chlorination or the equivalent.
3	Not more than 5,000/100 ml. and this MPN exceeded in not more than 20% of samples.	Rapid sand filtration, including coagulation, or its equivalent, plus continuous chlorination.
4	MPN greater than 5,000/100 ml. in more than 20% of samples and not exceeding 20,000/100 ml. in more than five per cent of the samples.	Auxiliary treatment such as 30 to 90 days storage, presettling, prechlorination or equivalent, plus complete filtration and chlorination.
5	MPN greater than 20,000/ml.	Prolonged storage or equivalent to bring within the limits of Groups 1 to 4.

[1] Anon. (1946).

A thorough course of treatment sufficient for making potable a raw water of low quality might involve: (1) sedimentation in a large impounding reservoir; (2) coagulation (the formation of a voluminous flocculent precipitate by adding iron or aluminum salts); (3) settling out the bulk of the precipitate in a reservoir or tank; (4) filtering the decantate through specially prepared beds of sand and gravel to remove residual precipitate and other impurities; (5) filtering through activated charcoal which will absorb many dissolved substances affecting the color, odor, and taste; and (6) chlorination. In some cases, the sequence of these operations may be changed, some of them may be omitted, and other steps may be added.

SEDIMENTATION

Plain sedimentation is precipitation of relatively large particles which occurs in more or less quiescent bodies of water. It necessitates retention of the water in a basin so particles can settle as a result of gravitational forces. As the sole method of purification, it is suitable only for relatively pure water containing undesirable amounts of suspended solid, but it can be an effective method of removing some of the grosser impurities from other raw water supplies. Table 10 compares the purifying effect of sedimentation and other procedures.

When a surface water supply is impounded in a reservoir which is sufficiently large, a satisfactory degree of sedimentation is often obtained over periods of weeks or months, and other treatment is not needed for clarification. Due to the large volume of the storage area involved however, it is often not economical to construct reservoirs which can take full advantage of natural sedimentation. But some sedimentation can be expected in almost all surface water systems. The sedimentation which occurs after coagulation is usually called "settling" to distinguish it from the simpler operation.

TABLE 10

COMPARATIVE PURIFYING EFFECTS OF SEDIMENTATION AND OTHER PROCEDURES[1]

	Location of Water Supply	
	Toledo, Ohio	Niagara Falls, N. Y.
Turbidity		
Raw water	185	21
Removed[2] by:		
Plain sedimentation	. . .	. . .
Coagulation	87.0	42.2
Bacteria		
Raw water	14,100	9,960
Removed[2] by:		
Plain sedimentation	. . .	. . .
Coagulation	80.0	50.1
Filtration	94.2	85.6
Chlorination	96.9	99.86
E. coli Index		
Raw water	1,650	8,290
Removed[2] by:		
Coagulation	81.9	64.9
Filtration	98.8	99.02
Chlorination	99.73	99.985

[1] Adapted from Anon. (1927).
[2] "Removed" is the percentage of the difference between the number in the treated water and the number in the raw water.

Sedimentation may economically precede coagulation, settling, and filtration if the water contains large amounts of coarse sediment and the water composition is correct for allowing the precipitation of the sediment. Often water with a high content of suspended matter is easier to clarify by plain sedimentation than is water containing only small amounts.

Sedimentation theory is based on Stoke's law which relates the rate of fall of a sphere in a liquid to viscosity of the fluid, size of the particle, and densities of the particle and the fluid. According to this law, when a small sphere falls under the action of gravity through a viscous medium it ultimately acquires a constant velocity, V, described by the equation

$$V = \frac{2\ ga^2\ (d_1 - d_2)}{9\eta}$$

where a is the radius of the sphere, d_1 and d_2 the densities of the sphere and the medium respectively, g is the acceleration of gravity, and η the coefficient of viscosity. *V* will be in centimeters per second if *g* is in centimeters per second square, *a* in centimeters, d_1 and d_2 in cubic centimeters, and η in dyne-second per square centimeters or poises.

The situation is not quite so simple in contaminated natural waters. Conditions tending to suspend and to prevent coalescence of relatively large particles of sediment vary greatly in different samples and can be determined only by experimentation, while convection currents modify the behavior to different degrees in different reservoirs.

Sedimentation basins can be operated on a continuous-flow method or on an intermittent (fill-and-flow) method. In continuous-flow operation, water is added and withdrawn at the same time and the same rate as it proceeds with a low velocity through one or more reservoirs in which sedimentation takes place. The intermittent method requires that water be let into a basin where it can remain quiescent during the period of subsidence. Clear water is drawn off when required and the basin then refilled with raw water. Preliminary coagulation is sometimes used in connection with sedimentation.

There are two disadvantages of the intermittent technique. Effective precipitation of fine particles does not begin until the fill is completed because of the turbulence created at the inlets. In addition, a loss of head results because the highest level of the water in the subsequent operation must be as low as the highest point from which the water is drawn off. Cleaning can be done by emptying the basin and flushing or scraping away the accumulated sediment. Mechanical devices which remove sludge without interrupting the operation are available.

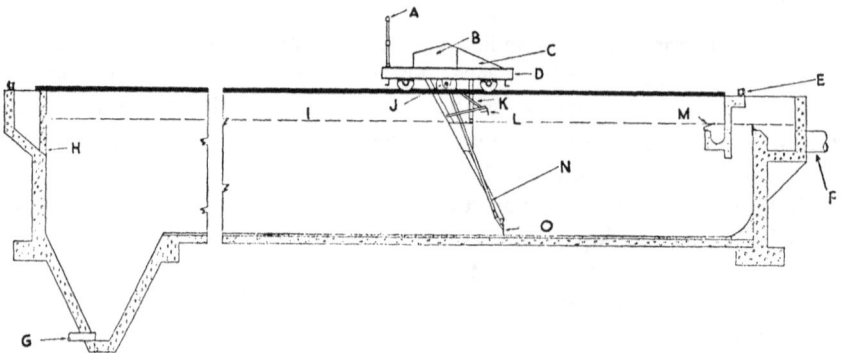

Courtesy of the Hardinge Co.

FIG. 8. PRINCIPLE OF OPERATION OF A RECTANGULAR CLARIFIER OR SETTLING TANK

The mechanism illustrated is traveling to the left, with the scraper (O) in lowered position, moving settled solids toward the sludge hopper. The skimmer (L) is in raised position but will be lowered on the return stroke.

Square or rectangular sedimentation basins (see Fig. 8) are more common than other shapes for the obvious reason that they are more economical to construct. Several basins may be connected in series. However, circular tanks with revolving scrapers for continuous removal of sludge are also used, in which case raw water is admitted at the center and drawn off from a peripheral trough called a launder (see Fig. 9). Spiral flow basins in which water enters tangentially have been shown to have very high efficiencies but may be more expensive and harder to control. The same effect may be achieved by installing baffles or partitions in a single large basin, so that turbulence from wind action or other causes does not mix turbid water with clearer fractions. Depths of horizontal flow basins normally range from 12 to 20 ft., averaging about 16 ft., including the allowance for sludge storage. Vertical flow basins may be as deep as 25 ft. The period of detention, defined as the volume of the basin divided by the volumetric rate of flow through it, varies in practice from a fraction of an hour to many days. Most well-designed installations have a detention period of six to eight hours—less if coagulation is used. Details of construction may be found in Babbitt *et al.* (1962), Bramer and Hoak (1962), Fair and Geyer (1954), Fair *et al.* (1927), Hardenbergh (1952), Nordell (1961), and Steel (1953).

Dimensions of suspended particles are so diverse, turbulence of flow so difficult to analyze, and ancillary factors so obscure, that theoretical treatment of the design of sedimentation basins has hitherto rarely been attempted and construction details were based on empirical principles derived from long experience. However, Bramer and Hoak (1962) presented a comprehensive set of equations which apparently permit design of sedimentation basins in the manner of the unit operation concept. Both approximate and more precise correlations applicable to particles either lighter or heavier than water were given. These correlations were substantiated by parallel operations of models of a steel rolling-mill scale pit in which the scale criteria predicted proved to be valid. Model performance comparisons were made on two bases: sedimentation performance and patterns of deposited solids.

COAGULATION AND SETTLING

Plain sedimentation removes at best only gross particles. It will not, of course, cause any reduction in colloidal substances which are primarily responsible for color and which contribute to turbidity. It has little or no effect on microscopic living organisms present, although concomitant oxidation and exposure to sunlight may cause a decline in numbers. Many years ago, it was found that certain voluminous flocculent precipitates, when formed *in situ*, carried down many colloidal components and effect-

ed a purification which it was not possible to obtain by plain sedimentation or in any other economically practicable way.   Now coagulation, as it
is called, is a very common step in water treatment.

The most effective, or at least the most popular, precipitates in water
engineering practice are ferric hydroxide and aluminum hydroxide.

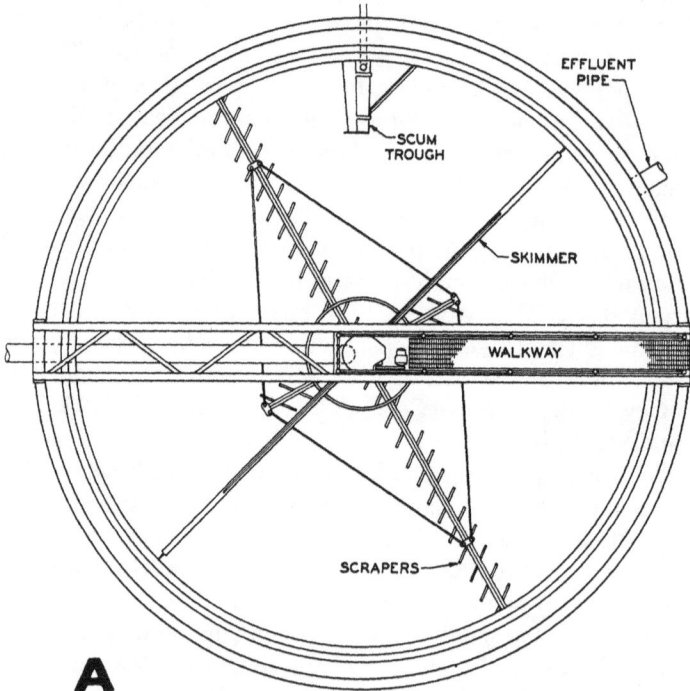

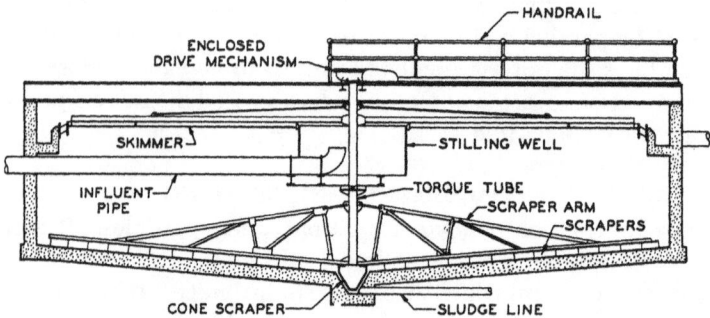

FIG. 9.   TWO TYPES OF CIRCULAR CLARIFIERS OR SETTLING
TANKS

Above—center feed type.   Facing page—under feed type

Perhaps these can be considered more accurately as colloidal oxide sols. They are formed by adding solutions of the sulfates of aluminum or iron to properly conditioned water. The trivalent cations are the critical operants. Turbidity and organic color in water are negatively charged. Mutual coagulation thus occurs simultaneously with mechanical entrapment. The flocculent precipitates and associated impurities are removed by settling and filtration. The amount of aluminum or ferric ion added to the finished water is very small, although the sulfate content may be increased appreciably.

COAGULATION

Coagulation, settling, and filtration are the usual steps for removing (or reducing to an acceptable level) the color and turbidity of raw water. Taste- and odor-causing materials as well as microorganisms are also reduced to a greater or lesser degree by these procedures. The purification

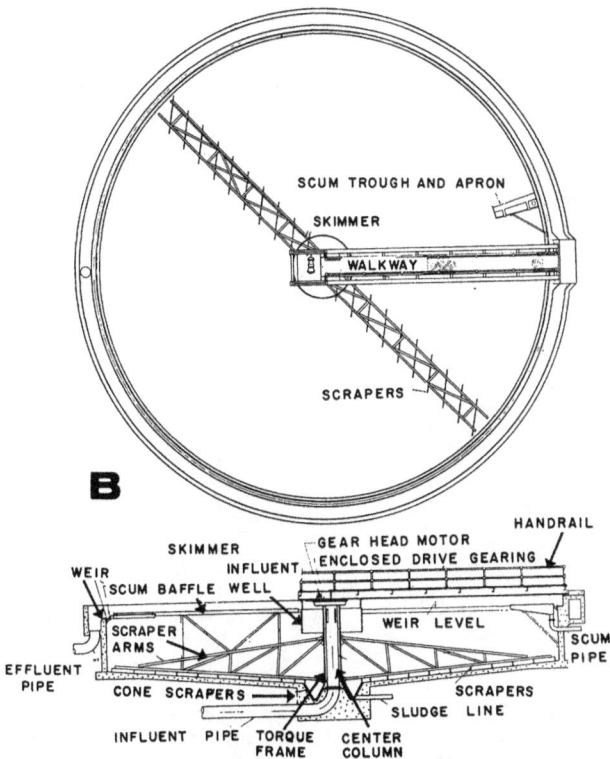

Courtesy of the Hardinge Co.

of water by forming voluminous flocculent precipitates *in situ* has been practiced for many years. Evidently the technique was first attempted because the precipitates resemble in some respects the "Schmutzdecke" or gelatinous coating of microbial origin that forms in the upper level of sand filters.

Coagulation is carried out for removal of suspended and colloidal impurities and certain dissolved materials which cannot be readily removed by simple sedimentation. A voluminous flocculent precipitate of the hydrated oxides of aluminum or iron (ferric) is formed in the water. The flocs collect the colloidal and particulate matter by direct and indirect mechanisms. The precipitate is then removed by settling and filtration. Taste and odor improvement depends upon degree of removal of organic

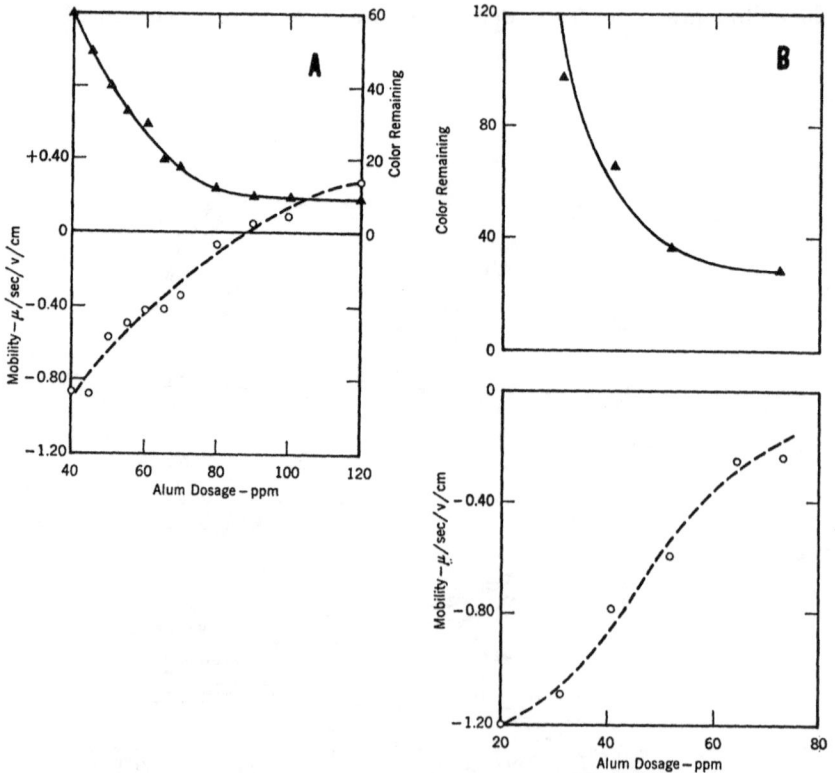

From Black and Willems (1961)

Fig. 10. Effect of Variations in Alum Dosage on the Coagulation and Color Removal in Two Samples of Water, the pH Being Held Constant at 4.85 (Left) or 5.25 (Right) by Addition of Hydrochloric Acid or Ca(OH)$_2$

colloids that is accomplished in conjunction with either color or turbidity removal.

The foundations of modern coagulation theory were laid down by Theriault and Clark (1923) and Miller (1923, 1924, 1925A, 1925B, and 1925C). In summary, these workers concluded that formation of a suitable floc is dependent upon: (1) the presence of a specific minimum concentration of ferric or aluminum ions (see Figs. 10 and 11); (2) an anion of strong

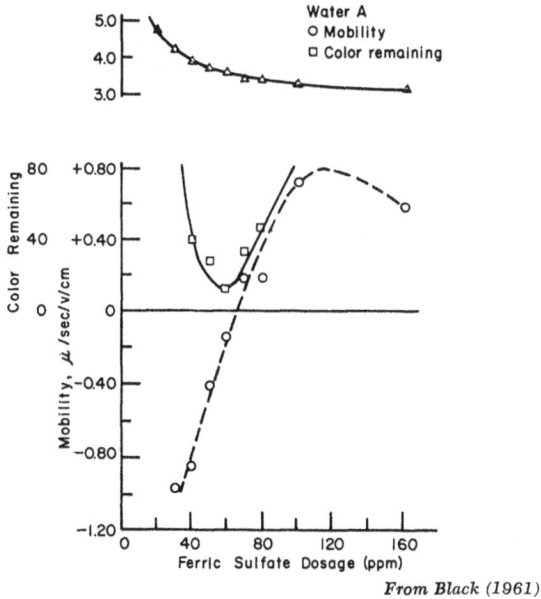

From Black (1961)

Fig. 11. The Effect of Dosage of Ferric Sulfate on Floc Mobility, Color Removal, and pH in Coagulation of a Colored Surface Water

coagulating potential; and (3) a favorable pH. The early publications were supplemented by many others which enlarged the area of knowledge of coagulation and made it more precise. The current state of coagulation theory has been discussed in detail by Black (1961 and 1963), who described the phenomenon as a complicated and incompletely understood process involving: (1) adsorption processes; (2) hydrogen-bonding effects; (3) reduction or reversal of the zeta-potential of the particle to be removed; and (4) electrostatic cross-linking involving polyvalent ions.

The zeta-potential which is regarded as one of the most important factors influencing coagulation effectiveness can be described briefly as the electrokinetic charge on suspended particles, especially on particles less than one micron in size. The sign of the zeta-potential can be reversed in most instances by adjustment of pH (see Fig. 12) but this may not be a practical procedure in water treatment. Bean *et al.* (1964), in a recent review, have interpreted the interaction between particles and coagulants on the basis of zeta-potential effects: (1) color removal requires a neutral zeta-potential when alum is used as the coagulant; (2) color removal requires a slightly positive zeta-potential when ferric sulfate is used as the

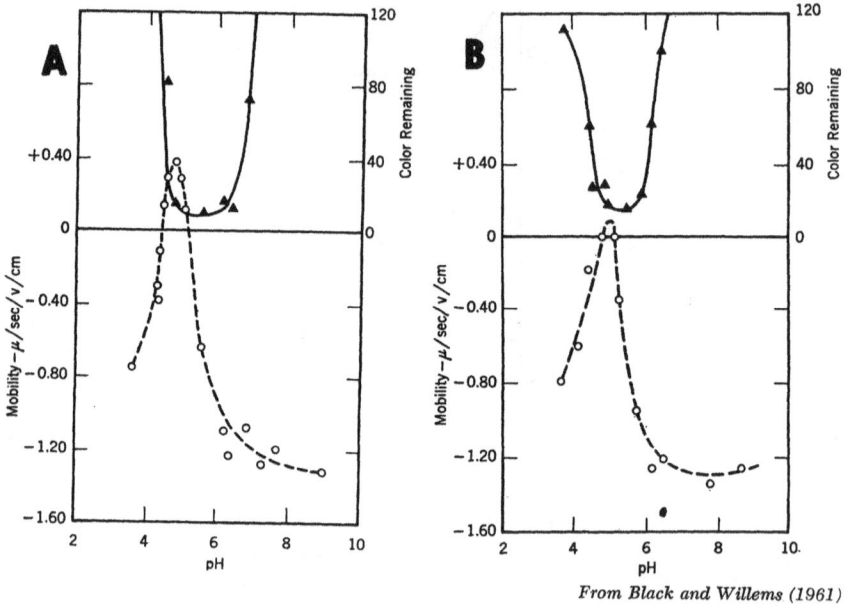

From Black and Willems (1961)

FIG. 12. EFFECT OF VARIATIONS IN pH ON COAGULATION AND COLOR REMOVAL IN TWO SAMPLES OF WATER, THE ALUM DOSAGE BEING HELD CONSTANT AT 60 P.P.M.

Left—water which was low in total solids and had an alkalinity of 5.0 p.p.m. Right—water which contained three times as much total solids as A, and had an alkalinity of 19 p.p.m.

coagulant; and (3) turbidity removal requires a somewhat negative zeta-potential. These authors believe that development of a method for automatically measuring zeta-potential could open the way for completely automatic control of coagulant dosage.

Clay turbidity and organic color are among the impurities affecting the zone of rapid floc formation. Both are present as negatively charged colloids at the usual pH of raw water. Black and Hannah (1961) found that the zeta-potential of clay particles depends on pH and alum dosage. Amounts of alum equal to several times the base-exchange capacity of the clay suspension were required to neutralize the particles' charge. Clarification was best in the pH range of 7.5 to 8.5 where particles were negative rather than at pH values where the particle charge had been neutralized. They tested eight different coagulant aids with alum and found that all could be made to either improve or inhibit coagulation by proper selection of dosages of alum and acid.

One characteristic of hydrated oxides of aluminum and ferric iron which makes them particularly suitable for water purification is the very low solids content of the floc. This makes it possible to treat large volumes of water with relatively small amounts of coagulating chemicals. Since the chemicals themselves are relatively inexpensive, especially in the partially purified form in which they are used, the cost factor is reduced to acceptable levels. Various forms and combinations of coagulant chemicals have been introduced over the years, each with some real or purported advantage over previous types.

According to Davis (1964), a satisfactory coagulant achieves two objectives. It will (1) neutralize the charge on the colloidal particles in the water to the point where they will coalesce into particles of sufficient size to settle and (2) yield a highly adsorbent hydrous oxide floc capable of removing finely divided materials as it settles or is filtered out. Color and turbidity are usually present as negatively charged colloidal particles and for this reason satisfactory coagulants are found to possess a positive charge.

Babbitt *et al.* (1962) list the following chemicals as being useful in coagulation operations:

(1) Aluminum sulfate, alum, filter alum. (2) Ferrous sulfate (copperas) with calcium hydroxide. (3) Activated alum—this is a proprietary acid-treated insoluble silicate containing 18.74% Al_2O_3 and less water of crystallization than filter alum. It forms a tough floc and aids in the coagulation of waters low in turbidity. (4) Ammonia alum $[Al_2(SO_4)_3(NH_4)_2 \cdot 24H_2O]$. Used as a coagulant in pressure filtration, since its solubility is about one-twentieth that of filter alum. It contains three to seven per cent of ammonia by weight. (5) Black alum is a filter alum containing two to five per cent of activated carbon. It is used to combine coagulation with removal of taste and odor. (6) Chlorinated copperas, a mixture of ferric chloride and ferric sulfate. Some consider it more effective in coagulation than ferrous sulfate. It is especially useful in conjunction with prechlorination but is not widely used. (7) Sodium aluminate ($Na_2Al_2O_4$), which may be used at a concentration of about 0.2 grain/gal. with alum in the coagulation of cold water. It is widely used in the lime-soda-water softening process to coagulate calcium carbonate and magnesium hydroxide.

Alum is the coagulant used by most municipal treatment plants. The formula of the solid substance is generally given as $Al_2(SO_4)_3 \cdot 18 H_2O$ but the commercial product often contains about 14 moles of water or slightly less. The reaction of this substance in contact with water is summarized in the following equations:

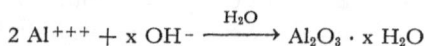

$$Al_2(SO_4)_3 \cdot 14 H_2O \longrightarrow 2 Al^{+++} + 3 SO_4^- + 14 H_2O$$

$$2 Al^{+++} + x OH^- \xrightarrow{H_2O} Al_2O_3 \cdot x H_2O$$

The hydrated oxides aggregate or polymerize into loose, small clumps, ideally about the size of a pinhead. Calcium sulfate is formed by reaction with the natural alkalinity of the water. If appreciable amounts of potassium or sodium ions are present, the alum floc may be reduced to almost colloidal size. At these dimensions, it is mostly ineffective in purification since the precipitate will not settle and may pass through the sand filters. Early publications showed that aluminum ion exerts its greatest effect on the zeta-potential of colloidal particles at pH 5.2. However, optimum pH for coagulation in most waters is near 7.0 or even very slightly on the acid side of neutrality. In some cases it may be as high as 8.0 or as low as 3.8 for highly colored soft waters. If necessary, pH may be adjusted upward by adding caustic soda, lime, and soda ash, or it may be lowered by adding carbon dioxide or sulfuric acid.

Iron salts such as ferrous sulfate are said to produce flocs that are faster-forming, denser, quicker-settling, and less easily broken up than alum flocs. These advantages are particularly evident at low temperatures. The ferrous ion is oxidized to the ferric state in the floc formation reaction:

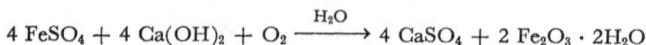

$$4 FeSO_4 + 4 Ca(OH)_2 + O_2 \xrightarrow{H_2O} 4 CaSO_4 + 2 Fe_2O_3 \cdot 2H_2O$$

The exact amount of water in the hydrated floc is, of course, indefinite. Bicarbonate alkalinity inhibits the reaction and if present, it must be converted to carbonate alkalinity, as by the addition of lime:

$$Ca(OH)_2 + HCO_3^- \longrightarrow CaCO_3 + H_2O + OH^-$$

Black *et al.* (1963) studied the stoichiometry of the coagulation of color-causing organic compounds with ferric sulfate and developed methods for determining optimum conditions of treatment. Two variables which must be controlled are coagulation pH and ferric sulfate dose. Both can be predicted with considerable accuracy from the raw water color alone.

In order for coagulation to be effective in purification, the water must contain nuclei approximately one millimicron in size.

The effectiveness of coagulation in relatively clear waters or in other difficult situations may be improved through the use of so-called coagu-

lant aids which supply nuclei of optimum dimensions and electric properties. Bentonite, fuller's earth, and other adsorptive clays, as well as activated silica and synthetic polyelectrolytes, are available for this purpose.
They are not themselves coagulants and are used with alum or similarly
acting chemicals. Their adsorptive powers often enhance the effect of
coagulants in taste and odor removal.

Different clays vary widely in their property as coagulant acids and
often the only reliable test is actual use in the given situation.

Reliable means for proportioning the coagulants and adjuncts must be
available for efficient and economical water purification. A simple and
reliable method for metering these chemicals is to prepare solutions of
known strength in independent mixing tanks, a duplicate set of tanks being
used for metering required during the mixing operation. From the active
set of tanks, the solutions are pumped or otherwise conveyed to small ori-
fice- or dosing-tanks in which the liquids are maintained at a constant
level, usually by applying an excess and allowing the surplus to overflow
across a baffle and return to the mixing tank. From the dosing tank, the
solution is fed through an orifice of known capacity discharging freely into
another receptacle or an open pipe. Delivery rate is adjusted by varying
the number or size of orifices or the head. Accurate dosage requires a
knowledge of the rate of flow of the water supply at the point of application but this information can be obtained easily enough from Venturi
measurements or the like.

After the raw water has been mixed with the coagulants, coagulant aids,
and other necessary chemicals (see Fig. 13), it flows into a reservoir
where turbulence is kept at a minimum so that the floc can settle to the
bottom. Of the various designs of settling basins and tanks which have

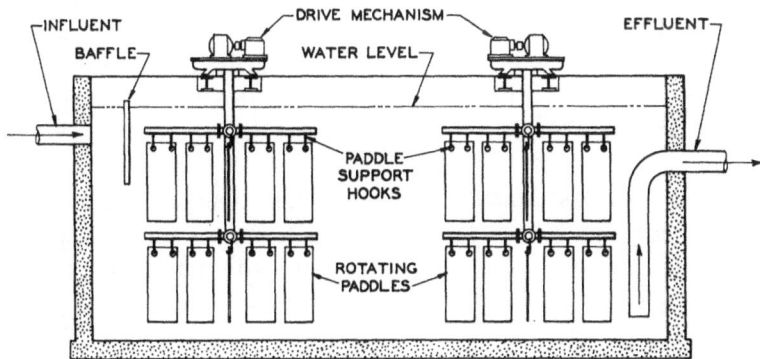

Courtesy of the Hardinge Co.

FIG. 13.   CROSS-SECTION OF A FLOCCULATING UNIT, SHOWING DETAILS OF
PADDLE MECHANISMS

been used, the suspended solids-contact or sludge-blanket units are the
most common today. In these, floc formation occurs while the water is
flowing downward through an inner section of the basin. The water then
rises and is filtered through a suspended blanket of previously formed
precipitates in an outer section. After it emerges from the upper surface
of the suspended layer or blanket it is collected at the drawoffs and passed
to the rapid sand filters.

The sludge can be removed from the bottom of the settling basin by
drawing it off through valved outlets in the bottom, which is either sloped
toward a single outlet or formed into a series of hoppers, each of the latter
having an outlet. The pressure of the overlying water will force much of
the sludge through the ports, but the final cleaning step will require flush
water as well as manual- or machine-scraping of the bottom. Continuously operated revolving scrapers are of value and are often included in high
capacity settling tanks of newer design.

Practice has shown that the ratio of surface area to volume of the basin
is a controlling factor in design. In fact, surface loading appears to be the
most important criterion of sedimentation efficiency when the horizontal
velocities in the basin are approximately uniform. Too great a surface
area in relation to tank volume is conducive to the creation of currents by
the wind. The American Water Works Association recommends that the
load per square foot of surface area of a settling basin be between 300 and
4,000 gal./sq.ft/day for granular solids, 800 to 2,000 for amorphous, slow-
settling solids, and 1,000 to 1,200 for flocculent material. For vertical
settling calculations, an average of two to three feet per hour is estimated
for alum-settling tanks and 3 to 6 feet per hour for lime-softening floc.

The rate of solids fall is inversely related to the velocity of flow. Flow
should not be greater than one foot per minute. The basin should be relatively narrow in comparison to length in order to minimize cross-currents
and not over 40 ft. wide in any case. Use of baffles to lengthen the flow
paths is contra-indicated because of the eddy currents they set up. Covers are not often used and may contribute more disadvantages than advantages.

As is the case with sedimentation reservoirs, settling basins may be operated with continuous flow (see Fig. 14) or on an intermittent basis.
These different methods have about the same advantages and disadvantages as those described for the former application.

FILTRATION

Filtration as it is practiced in water purification is not a simple mechanical removal of particles by a sieving action. Particles much smaller than

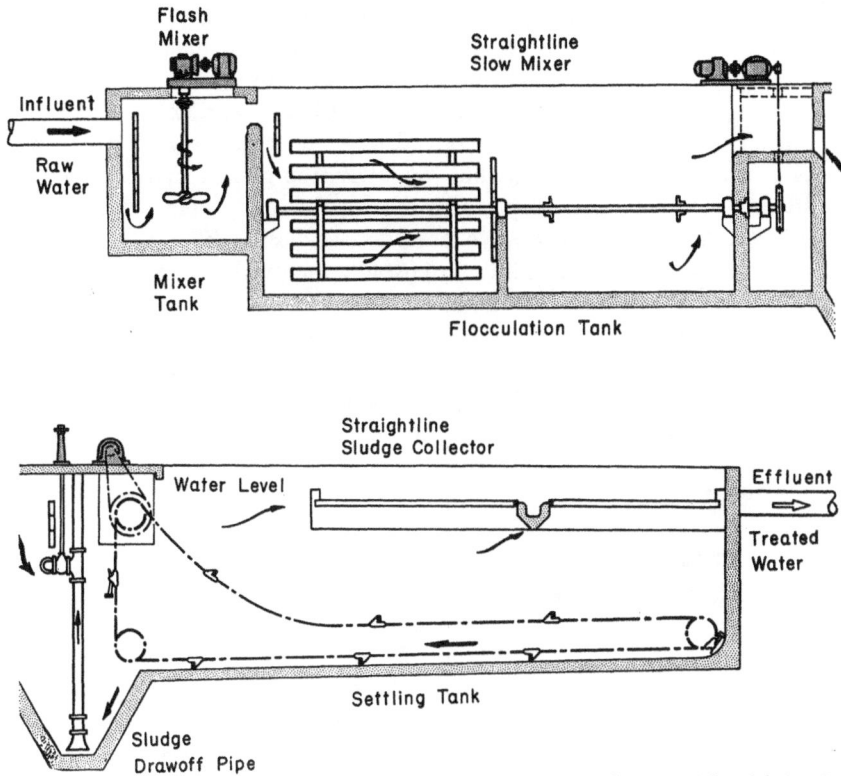

Courtesy of the Link-Belt Co.

FIG. 14. CHEMICAL FLOCCULATION EQUIPMENT

At left, a flash mixer provides the vigorous agitation necessary to mix the chemicals and raw water. Next, treated water flows into the flocculation tank where slow mixers promote maximum size floc formation. The water then goes to a settling tank where the floc coagulates, settles out, and is removed by a sludge collector.

the effective pore size of the filter are removed. Most filtration today is done by means of rapid sand filters, the older method of slow sand filtrations on the efficiency of filtration was published by Gobeck *et al.* (1964) economies of the new process. Flow through either type of sand filter is a gravity-powered process in all municipal installations of any size, but pressure filtration can be used in conjunction with aids such as diatomaceous earth, especially for industrial applications.

In the older process of slow sand filtration, a considerable part of the purification was due to a gelatinous coating, the Schmutzdecke, which built up around the sand granules in the top few inches of the filter bed.

This material resulted from the activity of innocuous bacteria which settled into the filter bed and used various impurities as substrates. It took the place, at least in some respects, of the coagulum of aluminum and ferric hydroxides which made rapid sand filtration workable. Turbidity (but not color) can often be removed by slow sand filtration without a preliminary coagulation stage, but the latter step greatly increases the effectiveness of the filter, especially during the early stages of its use. Coagulation and settling always precede rapid sand filtration (see Fig. 15), and both color and turbidity are usually removed to a satisfactory degree by this sequence of operations.

Flow rates through gravity-type rapid sand filters are generally held to not more than two gallons per square foot per minute, but some municipalities find higher rates of 2.5 to 3.0 to be acceptable. The standard rate in industrial practice is 3.0 gal/min./sq. ft. (Nordell, 1961).

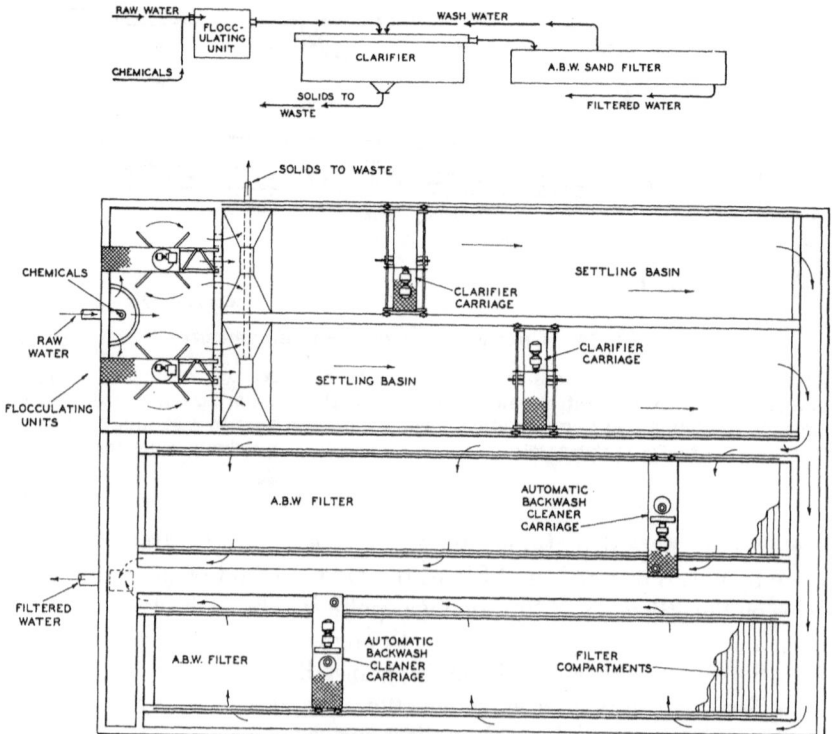

Courtesy of the Hardinge Co.

Fig. 15. Plan View and Diagram of Water of industrial Process Treating System, Complete with Flocculating Unit, Rectangular Clarifier, and Automatic Backwash Sand Filter

Shells of gravity-type rapid sand filters may be made of wood, concrete, or steel. Wood is no longer a common structural material, however, and concrete shells are the most common. The latter are usually square in the smaller examples and rectangular in the larger. Steel shells are nearly always cylindrical. The filtering medium may be sand or crushed and graded anthracite. The supporting layers under the filter bed can be composed of graded gravel or coarse anthracite. Special filter media may be required for certain purposes. For example, crushed and graded calcite is used in neutralizing filters to raise the pH of acidic corrosive waters to 7.2 or 7.3 and to stabilize treated waters. As discussed in a later section, activated carbon filters are useful in taste and odor removal but cannot replace sand filters because of the expense of carbon and its rapid inactivation by floc.

One of the most thorough studies of the effect of plant design modifications on the efficiency of filtration was published by Gobeck et al. (1964) who used pilot plants over a one and one-half year period to treat turbid water from the Little Miami River. They concluded: (1) a double-layered bed of coarse media made up of 18 in. of coal—effective size 1.05 mm.—on top of six inches of sand was able to remove as much or more turbidity, coliform bacteria, poliovirus, or powdered activated carbon as a bed of coal or sand alone and permitted the extension of filter runs beyond the usual life of the latter; (2) with proper coagulation ahead of the filter, a six gallon per minute per square foot filtration rate was as effective in removing all test particulates as rates of 4 or 2 gallons per minute per square foot; (3) adequate floc strength was more important in achieving clarification than a certain settleability when coarse media and high filtration rates were used, although this strength was frequently obtained only by the addition of 5 to 20 p.p.m. of activated silica as a coagulant aid or 0.05 to 0.2 p.p.m. of a synthetic polyelectrolyte as a filter aid; (4) when the water was relatively clear (less than 25 units of turbidity), flocculation and sedimentation steps of conventional treatment could be omitted if coarse media were placed on top of the sand; (5) for this particular river water, the inclusion of flocculation and limited sedimentation permitted longer filter runs and better water during winter and flood conditions; and (6) poliovirus in clear water was found to be more readily removed by beds of fresh, clean granular activated carbon than by sand, but carbon that was exhausted by a high loading of dissolved organic contaminants did not remove much more virus than a similar sized sand or coal bed.

The accumulation of strained out materials gradually clogs the filter bed and leads to a decreased flow and loss of head (pressure). The common method of cleaning rapid sand filters is by reverse flushing with treated water. This procedure accomplishes the desired result under all ordinary conditions and necessitates the consumption of only a small percentage of the water which has been filtered through the bed. Backwash requirements may vary from about 0.6 to 3.0% of the treated water, depending of course on design of the filter and extent to which the water is contaminated. The need for cleaning is determined by checking loss of head or time lag through the filter, or by measuring the amount of water which has been filtered, or by turbidity or color of effluent. Typical figures quoted in the literature are in the range of 10 to 18 hr. of run for a rapid sand filter, equivalent perhaps to 12 to 16 million gallons through a four million gallon filter (Anon. 1964). Wash would be applied to such a filter at the rate of 12 to 20 gal./sq. ft./min., or a total of about 80 to 90 thousand gallons. Loss of head between cleanings would be about 4 to 12 ft. in an average installation. After backwashing has been completed, the newly filtered water is piped to waste for a few minutes before the filter is connected to the service lines for a new cycle.

The cleaning of slow sand or secondary filter beds, which involves removing part of the surface layers, can be mechanized by track-laying machines moving on the sand, by tracked dumpers, by sand-skimming machines, or by trenching, according to Lewin (1961), who discussed the advantages and disadvantages of each of these methods. Smaller plants, including most industrial facilities, would not find these procedures economical because of the high initial capital outlay and would rely on manual methods.

An automatic valveless type of rapid sand filter, which has come into use in recent years, automates the cleaning process by backwashing with a predetermined amount of water when the loss of head reaches a set limit, say four to five feet. The system then filters the next batch of wash water and finally connects the subsequent flow to the service lines.

Although not absolutely necessary, it is desirable to install gravity filters in batteries of two or more units so that backwashing will not interrupt flow to the service lines (see Fig. 16).

Pressure type filters are made in both horizontal and vertical configurations. The vertical style (see Fig 17) ranges in size from 30 to 120 in. in diameter with capacities of 9 to 230 gal./min./unit at the standard flow rate of three gal./min./sq. ft. Horizontal filters are usually eight feet in diameter and from 10 to 25 ft. in length giving capacities of about 200 to 500 gal./min./sq. ft. Municipal water systems use such filters mostly in procedures for removal of iron and manganese from ground waters after a precipitate has been formed by aeration. Industrial utilization of pressure-filters is widespread because of the limited space required. These devices are cleaned by backwashing when the loss of head reaches 5 to 8 lb./sq. in. (see Fig 18).

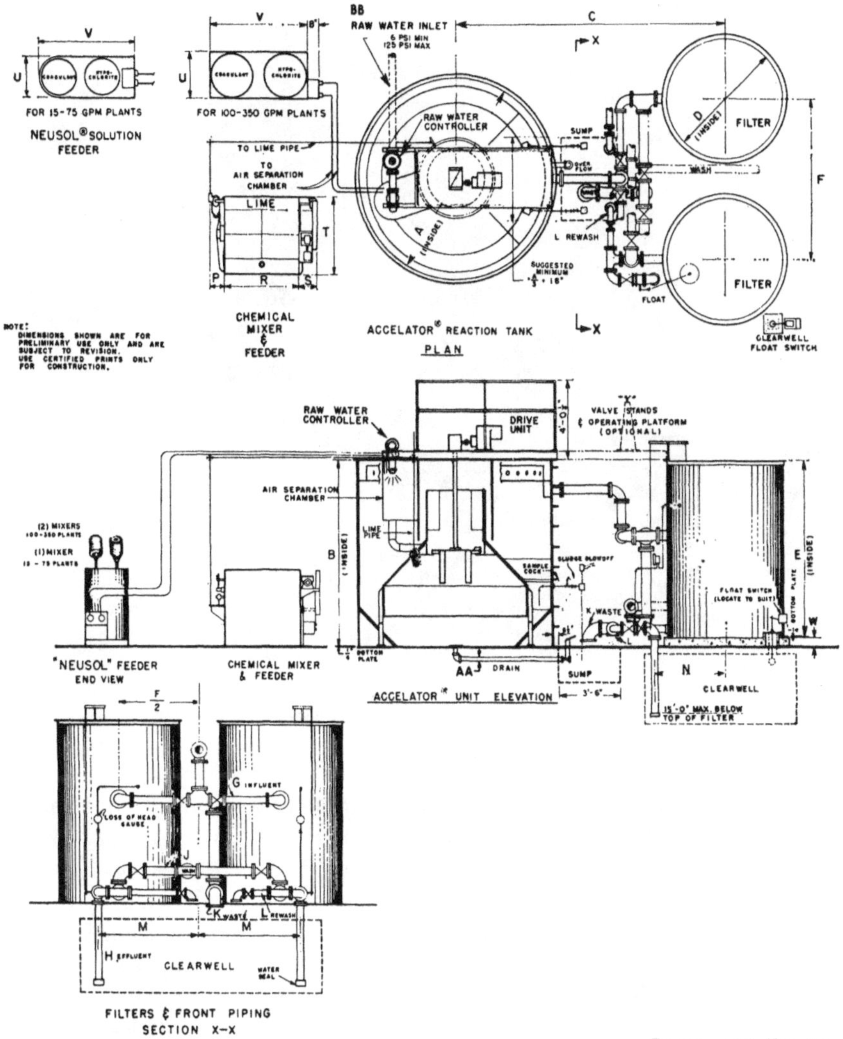

Courtesy of Infilco, Inc.

Fig. 16.   A Packaged Treatment Plant Using Gravity Filters

AERATION

Aeration in the broad sense might be considered to be any process in which the transfer of gases between a liquid and the atmosphere is facilitated. It is a unit process commonly employed in water- and waste-treatment plants to accomplish one or more of the following improvements: (1) Remove free carbon dioxide so as to increase the pH and thereby

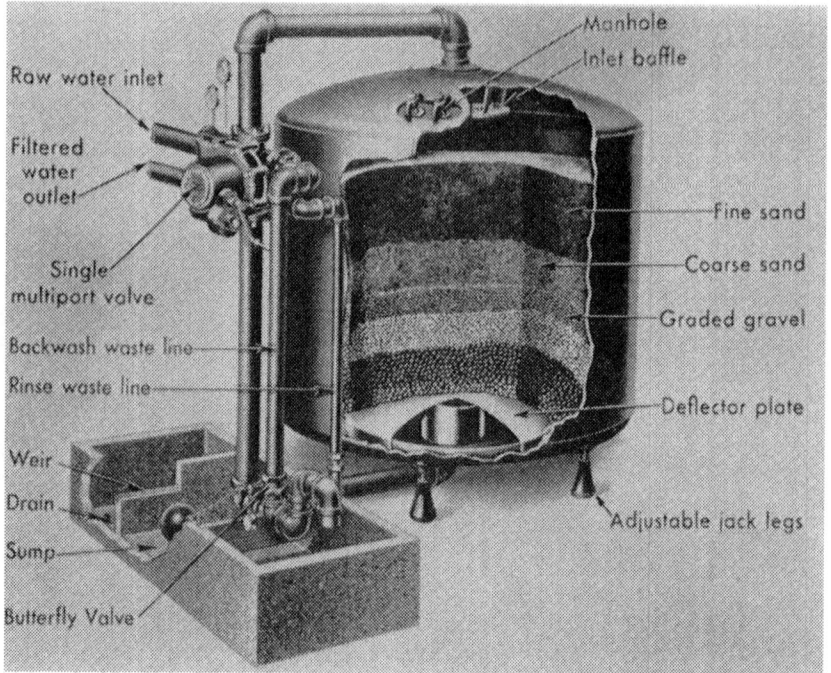

Courtesy of the Permutit Co.

FIG. 17. VERTICAL FILTER SHOWING ARRANGEMENT OF FILTER MEDIUM

reduce the corrosiveness of the water and decrease lime requirements in softening and coagulation steps. (2) Increase oxygen absorption to precipitate iron and manganese. (3) Promote the biochemical oxidation of organic matter in sewage digestion. (4) Deodorize by allowing hydrogen sulfide or volatile organic materials to distill off into the atmosphere. The fishy, grassy, or earthy odors often present in surface waters are mostly due to volatile oils originating from algae or other microorganisms. Aeration can remove much of the odor from these sources. (5) Remove excessive free residual chlorine. (6) Improve palatability by increasing the concentration of carbon dioxide in waters where this compound may be deficient. (7) Assist in various mechanical functions such as the distributing of coagulant or softening chemicals throughout the water, the backwashing of filter beds, etc.

Aeration can be accomplished by spraying the water, running it over louvers, or by any other method designed to increase the surface area of the liquid which is in contact with the atmosphere. Efficiency of any process is related to: (1) Time. (2) Temperature. An increase in tempera-

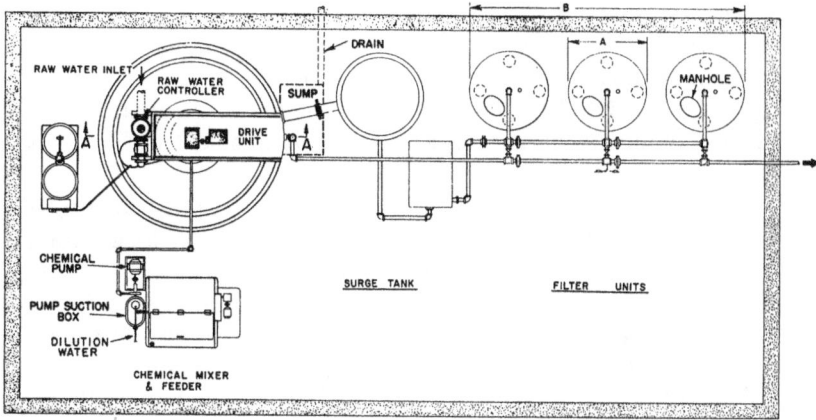

PLAN

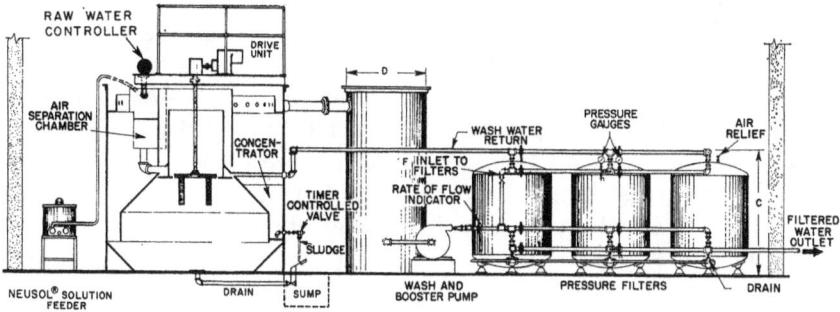

ELEVATION

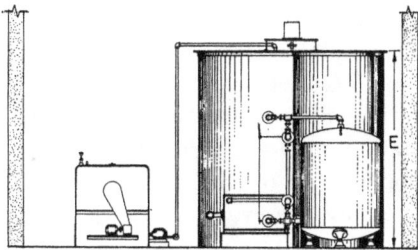

END ELEVATION

Courtesy of Infilco, Inc.

Fig. 18. A Packaged Treating Plant Using Pressure Filters

ture speeds the rate at which equilibrium is approached. (3) Surface area exposed. The greater the surface area exposed per unit time, the faster gas transfer occurs. (4) The rate of flow of air over the water surface. The transfer of gas is expedited by air turbulence or other means designed to bring greater volumes of atmosphere into contact with a given area of liquid surface. (5) The concentration of the gas in the water relative to its concentration in the atmosphere. The greater this ratio is, the more rapidly the gas is removed. (6) The solubility of the gas in the water, and the existence of combined forms such as hypochlorous acid and bicarbonate ion.

Most practical aerators are relatively simple in design. Some of the types which have been used are: (1) Baffle aerators. Baffles are used to provide turbulence of flow. (2) Cascade aerators. These are built in the form of steps or an inclined plane on which are arranged staggered projections to break up the surface of the water and bring more of it into contact with the air. (3) Coke tray aerators. Water is run through beds of coke arranged so as to allow air to enter freely. (4) Nozzle aerators. Water is squirted into the air as a fine spray issuing from a nozzle. (5) Diffusion aerators. Air is blown at low pressures through submerged porous plates or perforated pipes so that small air bubbles exit through the liquid. (6) Fountain aerators. These are made in the form of a tier of horizontal basins which decrease in diameter from bottom to top. Water is emitted from a nozzle above the top basin and descends to the bottom or collecting basin by flowing over the edge of each successively lower basin in turn. (7) Tray aerators. Water enters the top tray of a series of perforated trays of equal areas and falls as small droplets from one tray to another until it enters the bottom collecting tray. (8) Contact aerators. A biological oxidation unit consisting of several plates made of stone, cement-asbestos, or other corrosion resistant materials supported in a tank for sewage treatment. Air diffuses up and around the surfaces over which the waste material is flowing.

There are other kinds which are less common than the above.

Odors or tastes originating from industrial wastes or phenolic compounds are largely unaffected by aeration. Growth of algae may be increased by aeration under certain conditions of exposure of the water to sunlight and dust. It is obvious that aerators should be located where the atmosphere is relatively free of pollution so that odorous materials and other contaminants will not be entrained by the water in its passage through the device.

CENTRIFUGATION

Large centrifuges can be used to remove suspended solids from water supplies, waste liquid, etc. Generally speaking, they will not remove particles which would not settle out normally under the influence of gravity if given sufficient time. The effect is that of increasing the gravitational force hundreds or thousands of times so that the settling of particles is reduced to a few seconds. Extremely small particles or particles which have about the same density as the fluid in which they are suspended will not be removed either by sedimentation or centrifugation.

The utility of centrifuges is limited by the capacity and the initial cost of the equipment. Where space is so limited that settling basins are out of the question, centrifugation may very well provide an alternate solution to the problem of removing floc or other suspended solids. Combinations of centrifuges and pressure filters may have application in specialized circumstances. Centrifugation can be used to reduce the amount of waste material in an effluent, and is especially worthwhile if the retained material has some commercial value.

FLUORIDATION

Fluoridation will rarely, if ever, be practiced in food plants and the operation will not be covered in the present chapter for that reason. Readers interested in general information on methods for applying fluorides to drinking water should consult the work of Maier (1963).

DEODORIZATION WITH ACTIVE CARBON

If water completely free of foreign odors is required, as it is when water is to be used as an ingredient in certain foods and beverages, a final purification step in which it is passed through a bed of granular activated carbon may be required. Since water of such a high degree of excellence is not necessary for general consumption and because of the rather high cost of the process, only a few municipalities treat their supplies in this manner. Carbon in powdered form may be added with the coagulant chemicals and removed along with the floc in the settling basins and filter beds, or it may be used as a separate filter of the granular material.

Activated carbon removes by adsorption the compounds responsible for producing the off-tastes. This action is in contrast to the improving effect of chlorine, which deodorizes by changing the chemical nature of the objectionable substances. Activated carbon removes principally organic compounds and has little effect on the mineral composition. It will, however, remove residual chlorine. Since adsorption is a surface phenomenon, the effectiveness of any given sample of carbon will depend to some extent upon its surface area. Commercial activated carbons are normally rated in terms of square meters of surface area per gram. Baker (1964) described a direct method for evaluating the adsorptive power of activated carbons. His technique involved use of a gas-liquid chromatograph with a flame-ionization detector to determine the percentage removal of *n*-butane from an aqueous solution by the carbon in question.

Recovery of treatment chemicals is not common procedure in water purification technology, but in the case of activated carbon it is very desirable because of its relatively high cost. This is feasible only in those systems where granular carbon is used as a filter in a separate stage. Cycling systems involving thermal reactivation are frequently employed. Chemical regeneration seems to be less effective and more expensive than thermal processes. However, a plant in Culver City, Calif. obtains a 50% reactivation by passing a ten per cent solution of caustic soda through the spent beds.

Thermal reactivation generally involves (1) hydraulic pumping of saturated carbon from the filter to a dewatering device, (2) transferring the dewatered carbon to a multi-hearth furnace or rotary kiln; (3) quenching the reactivated carbon in water; and (4) pumping the slurry back to the filter (see Fig. 19). Flentje and Hager (1964A), as well as other authors,

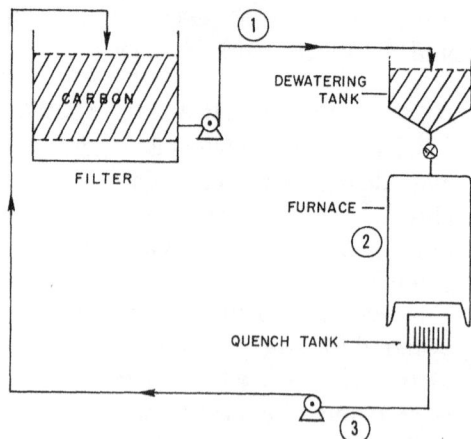

From Flentje and Hager (1964)

FIG. 19.   REACTIVATION PROCESS FOR GRANULAR CARBON

point out the necessity for using granular carbon with relatively high structural strength so that it will be capable of withstanding the repeated handling procedures without undergoing excessive breakdown. These authors indicate that loss during reactivation steps amounts to not more than five per cent per cycle. The cost of thermal reactivation, including fuel, labor, amortization, and carbon losses, is in the neighborhood of five cents per pound for medium to large plants.

A minimum bed depth of five feet and a maximum flow rate of one gallon per cubic foot per minute is recommended by Flentje and Hager (1964B). Greater efficiency has been obtained in some industrial installations by using beds 10 ft. or more deep with piping designed to allow countercurrent flow and series operation. Continuous removal of spent carbon and reintroduction of reactivated material are features of some of the newer plants.

SOFTENING

The "hardness" of water is essentially an indication of its ability to form insoluble precipitates with soaps although the term is used in water technology in reference to a specific index of the calcium and magnesium (sometimes iron and aluminum) content calculated as carbonates. Although variations in hardness within the normal ranges have no hygienic significance, they do affect potability and suitability of water for industrial purposes. Water containing more than 50 p.p.m. of the carbonates is considered to be "hard," while those with more than 100 p.p.m. qualify as "very hard." The ions responsible for hardness can affect the color, flavor, and texture of foods.

There are two main types of water-softening processes, (1) those in which the offending substances are precipitated by addition of appropriate chemicals and then removed by settling and filtration and (2) those in which calcium and magnesium are replaced by hydrogen or sodium ions in some sort of ion exchange process. The former category includes the cold-lime or lime-soda softening processes used by many municipalities. They are not as common in industrial installations where some modification of the zeolite or other ion-exchange process is more likely to be found.

The cold-lime process is the original water-softening technique, having been invented about 1841. It depends upon the precipitation of calcium carbonate and magnesium hydroxide which follows the addition of hydrated lime, $Ca(OH)_2$, in the form of lime-water or milk of lime (a mixture of dissolved and suspended hydrated lime) to water containing "bicarbonate hardness." In combination with soda ash, Na_2CO_3, lime is used to reduce both the bicarbonate hardness and non-carbonate hardness in the cold soda-lime process. One equivalent of hydrated lime is required to remove one unit of calcium bicarbonate hardness, while two equivalents are required to remove one unit of magnesium bicarbonate hardness. One equivalent of sodium carbonate suffices to precipitate one unit of calcium non-carbonate hardness. Both sodium carbonate and hydrated lime, one equivalent of each, are required to precipitate one unit of magnesium non-carbonate hardness. Since the precipitates yielded by these treatments are finely divided for the most part, a small amount of one of the conventional coagulants is usually added to facilitate settling and precipitation. The precipitated substances are somewhat soluble so there will be a residual hardness in water treated by the cold-lime or cold-soda-lime methods. The residual will be about 68 p.p.m. or four grains per gallon if hardness is reduced as much as possible without addition of excess chemicals.

The chief features in design of a softening plant relate to the equipment for preparing and introducing the chemicals, sedimentation basins for removing the larger particles of precipitate, and the final filtration step in which the finely divided precipitate is removed. There are three basic types of equipment in use today: (1) the suspended-solids contact or sludge-blanket type; (2) the older or conventional batch type; and (3) the "catalyst" or Spiractor type. Details can be found in Nordell's (1961) text, which includes a very complete discussion of water-softening theory and practice. In some cases, recovery of lime for re-use can be made with processes such as that shown in Fig. 20.

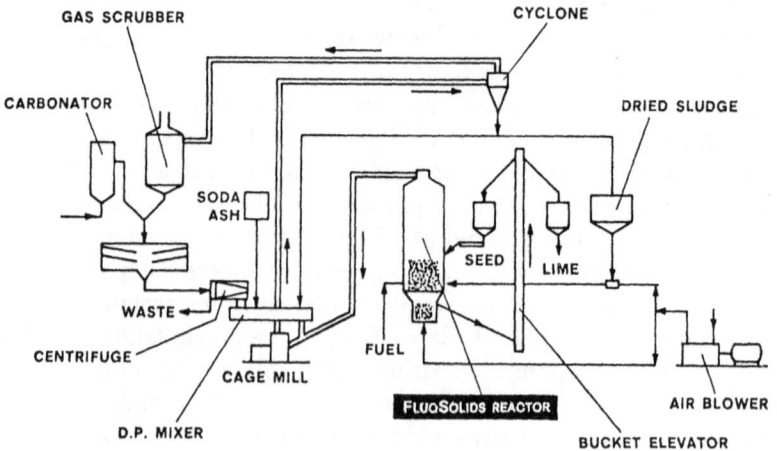

Courtesy of Dorr-Oliver, Inc.

FIG. 20. METHOD FOR RECOVERING LIME FROM WATER-SOFTENING SLUDGE

Many of the ion-exchange methods make use of natural or synthetic zeolites which are hydrous silicates. In the sodium cation exchange process, the water flows by gravity through a bed of the insoluble granular zeolite. The calcium and magnesium ions responsible for hardness displace an equivalent quantity of sodium ions from the mineral. Sodium ions do not contribute a "hardness" characteristic to water. The sodium content of the exchanger eventually becomes exhausted, of course, and is regenerated in three steps: (1) a strong current of water is passed up through the bed to loosen, clean, and regrade it; (2) a predetermined volume of concentrated sodium chloride solution is run through the bed, causing calcium and magnesium ions to be replaced by sodium ions; and (3) rinse water is passed through the bed to wash out the displaced calcium and magnesium and any sodium chloride solution remaining. These operations may be performed either automatically or manually.

As a rule, not all water going to consumers is treated by the ion-exchange process. Since almost all of the hardness is removed while a moderate degree of hardness can be tolerated, it is customary to mix treated effluent with unsoftened water to obtain an acceptable calcium and magnesium concentration.

The operating cost for removing non-carbonate hardness by the sodium cation exchange process is usually less than the expense for removing it by the cold-soda-lime process, but the latter procedure is less costly for bicarbonate softening. Sometimes the two methods are used in sequence to take advantage of the relatively low cost of the cold-soda-lime process for removing bicarbonate hardness and the correspondingly lower cost of the zeolite procedure for removing non-carbonate hardness.

Two bed exchangers can be used to remove both cations and anions from a supply, as shown diagrammatically in Fig. 21. Most processes will

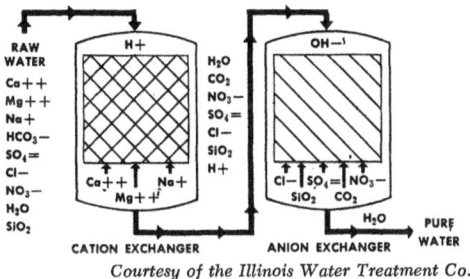

FIG. 21. DE-IONIZING WATER WITH A TWO BED EXCHANGER

Flow diagram

Courtesy of the Illinois Water Treatment Co.

not require water treated in this manner. If a very pure effluent is desired, equivalent to distilled water, the treated supply can be passed through a bed consisting of a mixture of cation- and anion-exchange resins. A large number of successive cation-anion exchanges result, many more than would be possible using separate beds of the two types of resins. Purity as high as 20,000 ohms/cm. can be obtained from a mixed bed unit. Special problems in regenerating the resins are encountered when the two types are mixed together. It is necessary to separate the resins by turbulent backwashing before the regenerant chemicals are applied. Remixing is accomplished by an aeration technique. The sequence of operations is shown in Fig. 22.

Ion-exchange resins require more expensive installations but they have

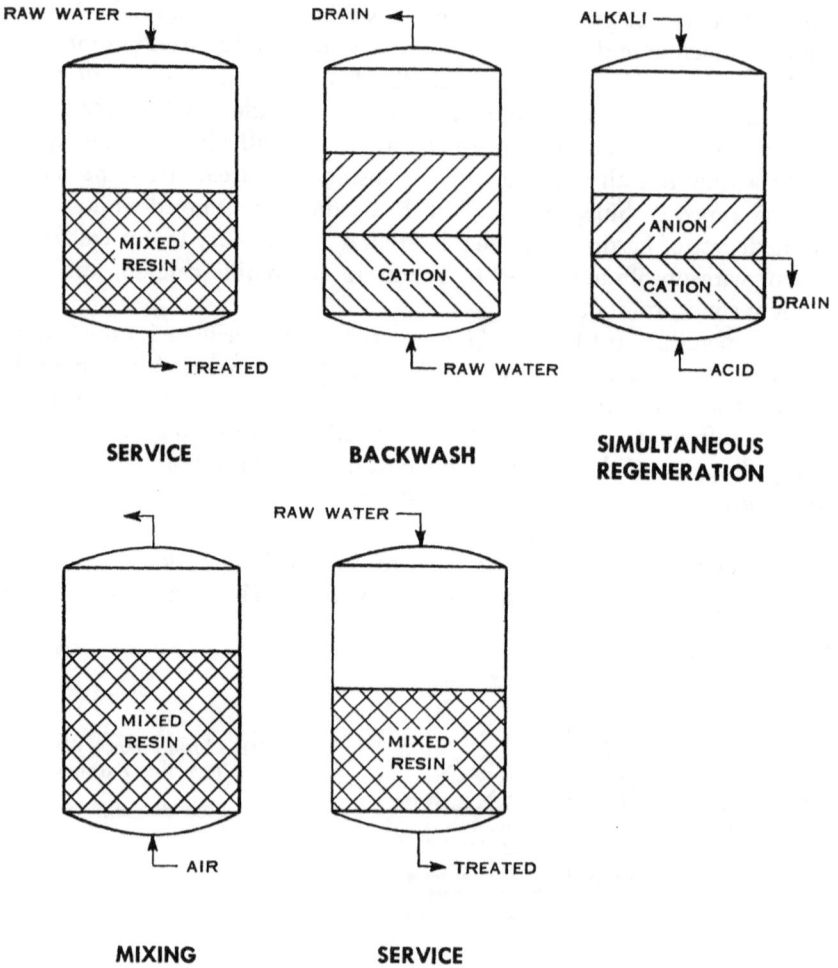

SERVICE   BACKWASH

SIMULTANEOUS
REGENERATION

MIXING   SERVICE

Courtesy of the Illinois Water Treatment Co.

FIG. 22.   STEPS IN THE REGENERATION OF A MIXED-BED DE-IONIZER

some advantages over the zeolite process, especially for relatively small scale industrial use. For one thing, calcium and magnesium can be replaced by hydrogen ions rather than sodium ions. This is a very real improvement in many processing applications.

IRON AND MANGANESE REMOVAL

In iron-bearing, deep-well waters which have the bicarbonate type of alkalinity, the iron is present as colorless and soluble ferrous bicarbonate,

$Fe(HCO_3)_2$. Iron in this form may be removed by oxidation to the insoluble ferric hydroxide followed by settling and filtration, or by some type of ion-exchange process. A manganese zeolite filter can be used to extract the iron if it is present in concentrations of about one p.p.m. or less. The sodium cation exchanger or zeolite water-softening processes will remove the iron along with the hardness. Ion-exchange resins in which the iron displaces hydrogen ions are used by some industries. Manganese is removed by analogous methods but a higher pH is required if the oxidation-precipitation technique is used. Oxidation can be effected by aeration or through such chemical oxidants as chlorine.

If the water is acidic, the iron and manganese may be present as the divalent sulfates $FeSO_4$ and $MnSO_4$. These can be removed by the successive operations of aeration, neutralization, settling, and filtration. Ferric hydroxide constitutes the major part of the iron content of the so-called "red waters." Since this compound is insoluble it may be removed by coagulation, settling, and filtration. Often the particle size is such that coagulation is not essential to the separation. Aeration may precede the other steps in order to bring other forms of iron to a more insoluble state.

Iron and manganese which are chelated by relatively high molecular weight organic compounds are frequently found in intensely colored waters. These, too, can be removed by aeration, coagulation, settling, and filtration.

SEQUENCE OF TREATMENTS

A published account by an anonymous author of the treatment applied in 1961 to Lake Michigan water by the city of Chicago provides an excellent summary of the sequence of operations involved in a typical municipal water supply system. The article is quoted and partially paraphrased below.

"Solutions of aluminum sulfate and iron sulfate are metered into the pipes and enough chlorine is added to convert to ferric salts the iron sulfate which is received largely as a ferrous salt. The iron and aluminum salts, reacting with the natural alkalinity of the raw water, slowly form hydroxides and begin to flocculate. As the insoluble hydroxides form, they heavily coat suspended matter, such as bacteria, plankton, and clay or sand particles, carrying a large part of it downward into the sediments which form in the settling basins where the water is next conducted.

"The iron sulfate solution mentioned above is a refined and concentrated pickling residue transported by tank truck from the steel mills on the south shore. The use of both iron and aluminum salts overcomes a problem of sensitivity to pH which each shows when used singly. Hydrolysis of the iron and aluminum salts causes the pH to drop toward 7. To offset corrosibility of the water to the iron mains, lime will be added to return the pH to slightly over 8. The addition of iron and aluminum salts have not resulted in increases of iron, aluminum, or calcium content of treated water at the South Plant. However, the sulfate content may rise to about 27 p.p.m. from the raw water concentration of 24 p.p.m.

"Not all iron and aluminum floc and encapsulated suspensoids are removed by settling. The last of the floc is caught on the surface of filter beds constructed of layers of fine sand on coarse sand on fine gravel on coarse gravel. Treated water, retaining a small proportion of the floc, will be poured over the filter beds and pass through by gravity flow until a gummy film of floc has formed on the surface. The water flow rate will drop from several to about two gallons per square foot per minute. The floc is then of the desired depth and permeability. Ninety-six filters, each with a daily capacity of 10 million gallons at normal flow rates, will be operating simultaneously. In such filters, after a day or two of operation, depending on contamination and demand, the rate of flow will begin to drop as more floc accumulates. Flow rates of each filter will be continuously and automatically registered and recorded. When the flow rate drops below a specified level, a backwash of filtered water will be automatically cycled to dislodge and remove the surface floc. Disengaged, used floc will be carried out into the lake and dumped. It would not be economical at the present time to attempt recovery of the iron and aluminum.

"During periods when odors are present, powdered activated carbon will be added with the coagulants to adsorb such odors. Dosages may vary from a few pounds of carbon to 10 pounds per million gallons of water. Carbon also reduces off-flavor.

"To the filtered water additional chlorine will be added to assure a complete kill of B. coli and substantially all other species of bacteria which might evade filtration. A residue of approximately 0.5 p.p.m. chlorine is expected when the water reaches the consumer an hour to several hours later. If, on examination at outlying pumping stations the chlorine level is found to have dropped, chlorine may be added there also. Since 1912 when the treatment of Chicago water was initiated, ammonia has sometimes been added to convert the chlorine into the more stable chloramine.

"Coffee and tea connoisseurs report a difference in the flavor of their beverages whether made with Chicago water containing a trace of chlorine or with the same water after having the chlorine removed by prolonged pre-boiling. They prefer the latter.

"In collaboration with the Public Health Service, the Bureau of Water began in 1954 to add fluoride in the form of a solution of fluosilicic acid to maintain a concentration of 0.9 to 1 p.p.m.

"Methods of analysis used by the Bureau are those of the American Water Works Association collaborating with the American Public Health Association and the Water Pollution Control Federation.

"The proportions of chemicals applied to water at the South District Plant in August 1961 were:

	p.p.m.
Aluminum sulfate (Calculated Al; factor 1.1)	1.4
Ferrous sulfate (Calculated Fe; factor 4.8)	2.6
Chlorine for oxidation	1.9
Lime, $Ca(OH)_2$	4.0
Carbon (variable)	1.8–18
Chlorine for sterilization	1.6
Anhydrous ammonia	0.2
Hydrofluosilicic acid (as F; factor 5.7)	0.9

"Filtered Lake Michigan water treated in the foregoing manner will be clear the year around; it will have a subthreshold level of residual chlorine, but sufficient to be hygienically safe; it will be more nearly neutral than raw lake water and less corrosive; it will be freed of most organic odor and flavor; it will contain dentally requisite fluoride, etc."

INDUSTRIAL TREATMENT PLANTS

Several firms have made available "packaged type" treatment plants in size ranges suitable for most industrial requirements. These are usually assembled from standardized components selected to yield water meeting the quantity and quality requirements of an individual situation. The "Accelapak" plants of Infilco, Inc. are examples of the package approach to industrial water treatment. A typical installation might include a raw water control, a slurry feeder for chemicals, coagulant and hypochlorite feeders, reaction tank and mixer, gravity or pressure filters, backwash pump and motor, activated carbon purifier, and automatic control devices. In some cases, one or more of these elements may be omitted while in other cases multiple units or special apparatus will have to be added to the basic plant.

The advantage of such installations is that many of the design uncertainties and some of the delays inherent to individualized layouts are decreased or eliminated. On the other hand, flexibility of design and adaptability to special problems may be less complete.

The quality of treated effluent can be expected to be equivalent to municipal water in all essential respects and to be satisfactory for most food plant use without additional processing. The effluent will often be superior to the available city water in taste and odor, especially if a charcoal filtration unit is included in the system. However, constant monitoring of water quality is essential. The quality control administrator of the factory should be assigned responsibility for maintaining the desired characteristics in the effluent and this may mean the hiring of additional technical personnel and the procuring of specialized laboratory equipment. These costs should be included in any analyses of the economic desirability of in-plant treatment facilities.

For the medium-sized plant having access to a reliable source of raw water, real economies can often be realized from the installation of a private treatment plant with the consequent elimination or reduction of metered water consumption. Opportunities for re-cycling of waste effluent may also exist.

BIBLIOGRAPHY

ANON. 1927. Studies of the effect of water purification processes. U. S. Public Health Serv. Public Health Bull. No. 172.

ANON. 1946. Manual of recommended water sanitation practice. U. S. Public Health Serv. Public Health Bull. No. 296.

ANON. 1951. Water Quality and Treatment. Second Edition. American Water Works Association, New York.

ANON. 1964. Symposium of filter practice. Water Works and Wastes Eng. 1, No. 3, 76, 78.

BABBITT, H. E., DOLAND, J. J., and CLEASBY, J. L. 1962. Water Supply Engineering. Sixth Edition. McGraw-Hill Book Co., New York.

BAKER, R. A. 1964. Chromatographic evaluation of activated carbon. J. Am. Water Works Assoc. 56, 92–98.

BEAN, E. L., CAMPBELL, S. J., and ANSPACH, F. R. 1964. Zeta-potential measurements in the control of coagulation chemical doses. J. Am. Water Works Assoc. 56, 214–224.

BETZ, W. H., and BETZ, L. D. 1950. Betz Handbook of Industrial Water Conditioning. Third Edition. W. H. and L. D. Betz, Philadelphia.

BLACK, A. P. 1961. Theory of coagulation. Water and Sewage Works 108, Reference Number, R192–194, 196, 198–199.

BLACK, A. P. 1963. Better coagulation processes for better waters. Water Works Eng. 116, 375–380.

BLACK, A. P., BALL, A. I., BLACK, A. L., BOUDET, R. A., and CAMPBELL, T. N. 1959. Effectiveness of polyelectrolyte coagulant aids in turbidity removal. J. Am. Water Works Assoc. 51, 247–263.

BLACK, A. P., and CHRISTMAN, R. F. 1961. Electrophoretic studies of sludge particles produced in lime-soda softening. J. Am. Water Works Assoc. 53, 737–747.

BLACK, A. P., and HANNAH, S. A. 1961. Electrophoretic studies of turbidity removal by coagulation with aluminum sulfate. J. Am. Water Works Assoc. 53, 438–452.

BLACK, A. P., SINGLEY, J. E., WHITTLE, G. P., and MAULDING, J. S. 1963. Stoichiometry of the coagulation of color-causing organic compounds with ferric sulfate. J. Am. Water Works Assoc. 55, 1347–1366.

BLACK, A. P., and WALTERS, J. V. 1964. Electrophoretic studies of turbidity removal with ferric sulfate. J. Am. Water Works Assoc. 56, 99–110.

BLACK, A. P., and WILLEMS, D. G. 1961. Electrophoretic studies of coagulation for removal of organic color. J. Am. Water Works Assoc. 53, 589–604.

BRAMER, H. C., and HOAK, R. D. 1962. Design criteria for sedimentation basins. Ind. Eng. Chem. Process Design and Development 1, 185–189.

BRANSBY-WILLIAMS, G. 1946. The Purification of Water Supplies. Second Edition. Chapman and Hall, London.

CAMP, T. R. 1964. Theory of water filtration. Proc. Am. Soc. Civil Engr. Sanitary Eng. Div. 90, No. SA4, 1–30.

CHAMOT, W., and STEWARD, B. 1963. Coagulation by sodium aluminate and its implications on the mechanism of coagulation. Am. Chem. Soc. Div. Water Waste Chem. Preprints 1963, 45–49.

DAVIS, W. B. 1964. A review of the theory of coagulation. Southwest Water Works J. 45, No. 10, 10–14.

FAIR, G. M. 1951. The hydraulics of rapid sand filters. J. Inst. Water Engrs. 5, 171–180.

FAIR, G. M., FLINN, A. F., WESTON, R. S., and BOGERT, C. L. 1927. Waterworks Handbook. Third Edition. McGraw-Hill Book Co., New York.

FAIR, G. M., and GEYER, J. C. 1954. Water Supply and Waste Disposal. John Wiley, New York.

FLENTJE, M. E., and HAGER, D. G. 1964A. Advances in taste and odor removal with granular carbon filters. Water and Sewage Works 111, 76–79.

FLENTJE, M. E., and HAGER, D. G. 1964B. Reevaluation of granular-carbon filters for odor and taste control. J. Am. Water Works Assoc. 56, 191–197.

FOXWORTHY, J. E., and GRAY, H. K. 1958. Removal of hydrogen sulfide in high concentrations from water. J. Am. Water Works Assoc. 50, 872–878.

GAINEY, P. L., and LORD, T. H. 1952. Microbiology of Water and Sewage. Prentice-Hall, Englewood Cliffs, N. J.

GOBECK, G. G., DOSTAL, K. A., and WOODWARD, R. L. 1964. Studies of modification in water filtration. J. Am. Water Works Assoc. 56, 198–213.

GRANSTROM, M. L., and SHEARER, S. D. 1958. Conductimetric control of coagulant dosage in treatment plants. J. Am. Water Works Assoc. 50, 410–416.

HARDENBERGH, W. A. 1952. Water Supply and Purification. Third Edition. International Textbook Co., Scranton, Penna.

HUDSON, J. E., JR. 1956. Factors affecting filtration rates. J. Am. Water Works Assoc. 48, 1138–1154.

JAMES, A. 1964. The bacteriology of trickling filters. J. Applied Bact. 27, 197–207.

LEWIN, J. 1961. Mechanization of slow sand and secondary filter bed cleaning. J. Inst. Water Engrs. 15, 15–46.

MAIER, F. J. 1963. Manual of Water Fluoridation Practice. McGraw-Hill Book Co., New York.

McCALLUM, G. E. 1959. Water supply protection activities of the U. S. Public Health Service. J. Am. Water Works Assoc. 51, 575–580.

MILLER, L. B. 1923. On the composition of the precipitate from partly alkalinized solutions. U. S. Public Health Serv. Public Health Repts. 38, 1995–2000.

MILLER, L. B. 1924. Adsorption by aluminum hydroxide considered as a solid solution phenomenon. U. S. Public Health Serv. Public Health Repts. 39, 1502–1516.

MILLER, L. B. 1925A. A study of the effect of anions upon the properties of "alum floc." U. S. Public Health Serv. Public Health Repts. 40, 351–367.

MILLER, L. B. 1925B. Notes on the clarification of colored waters. U. S. Public Health Serv. Public Health Repts. 40, 1472–1481.

MILLER, L. B. 1925C. Some properties of iron compounds and their relation to water clarification. U. S. Public Health Serv. Public Health Repts. 40, 1413–1419.

NORDELL, E. 1961. Water Treatment for Industrial and Other Uses. 2nd Edition. Reinhold Publishing Corp., New York.

PACKHAM, R. F. 1963. Some recent investigations into the mechanism of the coagulation process. Am. Chem. Soc. Div. Water Waste Chem. Preprints 1963, 40–44.

POWELL, S. T., and VON LOSSBERG, L. G. 1948. Removal of hydrogen sulfide from well water. J. Am. Water Works Assoc. 40, 1277–1290.

RIDDICK, T. M. 1963. The role of the zeta-potential in coagulation and dispersal. Am. Chem. Soc. Div. Water Waste Chem. Preprints 1963, 20–24.

ROBECK, G. G., DOSTAL, K. A., and WOODWARD, R. L. 1964. Studies of modifications in water filtration. J. Am. Water Works Assoc. 56, 198–213.

RUEHRWEIN, R. A., and WARD, D. W. 1952. Mechanism of clay aggregation by polyelectrolytes. Soil Sci. 73, 485–492.

RUTH, B. F. 1946. Correlating filtration theory with industrial practice. Ind. Eng. Chem. 38, 564–570.

SALVATO, J. A., JR. 1958. Environmental Sanitation. John Wiley and Sons, New York.

STEEL, E. W. 1953. Water Supply and Sewerage. McGraw-Hill Book Co., New York.

THERIAULT, E. J., and CLARK, W. M. 1923. An experimental study of the relation of H+ concentration to the formation of floc in alum solutions. U. S. Public Health Serv. Public Health Repts. 38, 181–190.

WELLS, S. W. 1954. Hydrogen sulfide problems of small water systems. J. Am. Water Works Assoc. 46, 160–170.

WELSH, G. B., and THOMAS, J. F. 1960. Significance of chemical limits in USPHS drinking water standards. J. Am. Water Works Assoc. 52, 289–300.

WILLEY, B. F., JENNINGS, H., and MUROSKI, F. 1964. Removal of hydrogen sulfide with potassium permanganate. J. Am. Water Works Assoc. 56, 475–479.

WOODWARD, R. L. 1964. Relation of raw-water quality to treatment plant design. J. Am. Water Works Assoc. 56, 432–440.

CHAPTER 6

Disinfection

INTRODUCTION

In industrial water treatment practice, disinfection or the destruction of disease-causing organisms can be considered to be virtually synonymous with chlorination. Various other disinfection methods such as heating, irradiation with ultraviolet rays, application of ultrasonic energy, and the addition of ozone, chlorine dioxide, lime, bromine, iodine, or colloidal silver have been advocated from time to time, and some of these are doubtless being used to a limited extent, but no other technique has proved to be more effective than, or as inexpensive as, chlorination for general water treatment purposes. The reliability of chlorine as a disinfectant for municipal water supplies is indicated by the data in Table 11.

An interesting historical account of the first attempts to introduce chlorination as a purification technique for water supplies can be found in the recent paper by Nesfield (1962). Hypochlorite of lime—bleaching powder—was first used extensively for water purification in 1908 both at the Chicago Stock Yards and at Jersey City. Between 1908 and 1915, it came to be used rather widely for the disinfection of public water supplies. Subsequently to the later date, it began to be replaced by chlorine gas which is particularly suitable for water disinfection because it is inexpensive, easy to apply, and has harmless end products. It is very effective in destroying bacteria, but a few other kinds of microorganisms are more resistant to ordinary levels of chlorination. Among the latter is *Entamoeba histolytica*, the causative agent of amoebic dysentery. Many viruses are markedly more resistant to chlorine than are bacteria such as *E. coli*. Among the more resistant viruses are the Coxsackie A2 virus, some at least of the polioviruses, and probably the causative agent of infectious hepatitis.

There have been several controversies regarding the mechanism by which chlorination kills microorganisms. Some of the hypotheses which have been advanced are:

(1) Chlorine reacts with water to form HOCl which then decomposes to yield the active agent, nascent oxygen. The latter oxidizes some part of the organic material of the bacteria. This view is no longer held.

(2) The free chlorine directly attacks reactive sites on the cell walls of the microorganisms. The explanation is consistent with the fact that the free element is a very strong oxidizing agent.

77

TABLE 11
EFFECTIVENESS OF CHLORINE IN DISINFECTING WATER SUPPLIES AS MEASURED BY COLIFORM REDUCTION[1]

Daily Range, MPN/100 Ml.	Coliform Density in the Raw Water — Frequency, Days				Chlorinated Water — Days on Which Coliform Positive Samples Were Detected[2]	
	Plant C-1	Plant C-2	Plant C-3	All Plants	Number	Per Cent
0 to 24	501	322	61	884	4[3]	0.5
25 to 49	...	172	21	193	0	0
50 to 99	...	59	0	59	0	0
100 to 240	40	91	15	146	2[4]	1.4
250 to 490	...	17	0	17	0	0
500 to 990	...	1	0	1	0	
1000	1	2	1	4	0	0

[1] From Walton (1961).
[2] At plant C-1, coliform bacteria (MPN of 2.2/100 ml.) were detected in two samples of chlorinated water examined on days for which raw water data were not available.
[3] Coliform densities 9.5, 5.1, 2.2, and 2.2; average 4.6/100 ml.
[4] Coliform densities 2.2 and 2.2; average 2.2/100 ml.

(3) The bacterial protoplasm is coagulated by the disinfectant.

(4) A key enzymatic process is inhibited by the chlorine. However, the action of the chemical appears to be quite non-specific and it is not necessary to postulate action on some particular enzyme to account for its lethal effect.

(5) The active form is hypochlorous acid. Such an explanation is consistent with the observed fluctuation of chlorine potency as the pH is varied.

Many authors have remarked on the small concentration of chlorine which will kill *E. coli*. However, as pointed out by Ingols *et al.* (1953), the ratio of the amount of chlorine in a bactericidal solution to the amount of cellular material with which it reacts is relatively large. In a sample calculation, these authors arrive at a figure of 250,000 organisms per microgram. One microgram of chlorine in a gram of solution would be equivalent to one part per million on a weight basis. Thus water containing 2.5×10^5 organisms per milliliter and one part per million of chlorine would have approximately equal amounts of the bactericide and bacteria. Most waters met with in practice are not this highly contaminated and the amount of chlorine applied is greater than one part per million so that the ratio of the latter to the former is much greater than unity.

Ingols *et al.* (1953) suggested that hypochlorous acid and chlorine dioxide are effective as bactericidal agents in water treatment because they form an irreversible product by reaction with sulfhydryl radicals on the bacterial proteins, inactivating enzymes, coagulating membranes, and disorganizing the cell in other ways.

As stated above, the effectiveness of chlorine is known to be related to the hydrogen ion concentration. About 150 times more chlorine must be applied at pH 10 than at pH 5 to get the same killing effect. The influence of pH variations may be somewhat obscured if there is much organic matter present since chlorine tends to react with this material in preference to bacterial cells. That is, organic matter and turbidity particles react with the chlorine to prevent the formation of hypochlorous acid and reduce the bactericidal effect. Iron, manganese, hydrogen sulfide, and nitrogenous compounds such as amines can also react with chlorine, reducing its effective concentration.

The formation and dissociation of hypochlorous acid is said to occcur in conformity to the following equation:

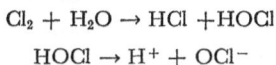

$$Cl_2 + H_2O \rightarrow HCl + HOCl$$

$$HOCl \rightarrow H^+ + OCl^-$$

Hypochlorous acid predominates at lower pH levels while the ionized form is predominant at the higher pH levels. This tendency holds good for bromine and iodine also, as shown in Fig. 23. The most effective form of bromine is also the hypohalous acid, as with chlorine, but iodine in the molecular form appears to be a more active bactericide than hypoiodous acid. The relative killing power of these three acids, as a function of concentration, is graphed in Fig. 24.

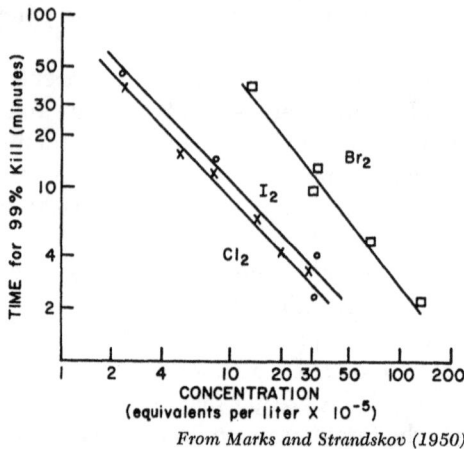

From Marks and Strandskov (1950)

FIG. 23. COMPARATIVE KILLING ACTION OF DIFFERENT CONCENTRATIONS OF CHLORINE, BROMINE, AND IODINE ACTING ON B. metiens SPORES AT pH 7 AND ROOM TEMPERATURE

The disinfection action of chlorine increases with increasing temperature, at least within the range of temperatures encountered in water treatment practice. At neutral pH, about $2^3/_4$ times as much chlorine is required at 40° as at 77°F. The contact time is also an important factor in securing adequate disinfection. Different organisms may vary widely in the rate at which they succumb to a given concentration of chlorine. The relative effectiveness of free chlorine at various pH's and temperatures in destroying viruses seems to parallel its fluctuations in lethality to bacteria.

Chlorine may be applied before (prechlorination), after (postchlorination), or between other forms of water treatment. It may also be applied at more than one point in the cycle, a practice known as rechlorination.

Chlorination may be accomplished by injection of the substance from cylinders containing 100, 150, or 2,000 lb. of the liquid, or by means of hypochlorinators which feed small quantities of sodium or calcium hypochlorite into the system. Use of the latter equipment is generally restricted to the smaller plants. Babbitt et al. (1963) list four possible bases for classifying chlorine feeders: (1) the state of the chlorine: is it a dry gas fed through a diffuser at the point of application or is it a solution of chlorine in a small amount of water? (2) The type of solution feeder: (a) hydraulic, (b) hydraulic-mechanical, or (c) mechanical. (3) The method of applying vacuum to the solution feed: (a) the chlorine is maintained under a vacuum throughout the apparatus, (b) a vacuum is maintained only when the eductor is in operation, or (c) pulsating conditions are used. (4) The type of control: (a) an operator controls the feed by manually setting a valve; (b) the metering device is connected to some other piece of apparatus which turns the chlorinator on or off; (c) the rate of feed is varied automatically in accordance with rate of water flow; and (d) the dosage is automatically and continuously controlled by sensors for residual chlorine or oxidation-reduction potential and flow.

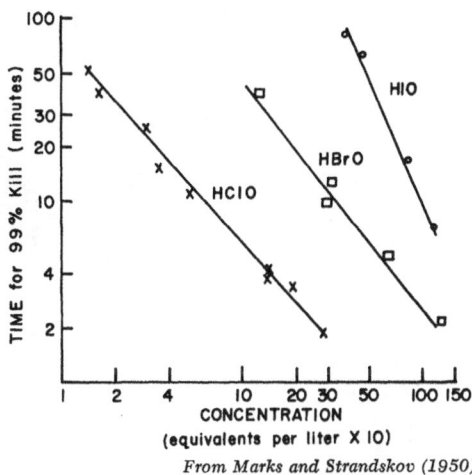

From Marks and Strandskov (1950)

FIG. 24. EFFECT OF CONCENTRATION OF HYPOHALOUS ACIDS ON KILLING TIME
B. metiens SPORES AT 77°F AND pH 6

Automatic control of chlorine application has been made possible by amperometric techniques for measuring residual chlorine. Such tests depend upon the ability of chlorine and iodine to generate a current proportional to the concentration in an amperometric cell. The calibration of such equipment is often done with a titration test in which the end-point is also determined by an amperometric device.

The point of application must be selected to allow thorough mixing of the chlorine with the water before it reaches the first user outlet. The chlorination apparatus and spare cylinders should be located in a separate gas-tight room which can be ventilated to an unoccupied area if leakage occurs. The door to the room should have a glass inspection panel. A gas mask furnished with a chlorine-absorbing cannister should be stored near the entrance to the room. The temperature of the cylinders should be cooler than the room in which they are stored but not less than 50°F. in order to prevent condensation of the substance in the lines (Salvato 1958).

Figure 25 illustrates a vacuum chlorinator manufactured by Capital Controls Company and shows the method of attachment to the ejector and the diffuser in the water line. The set-up as shown here is capable of feeding from 0.6 to 100 lb./day. When wall-mounted and hooked to a multiple-cylinder arrangement, it can feed up to 250 lb./day. The same manufacturer also supplies dual meter units which can apply pre- and post-chlorination in water or waste treatment.

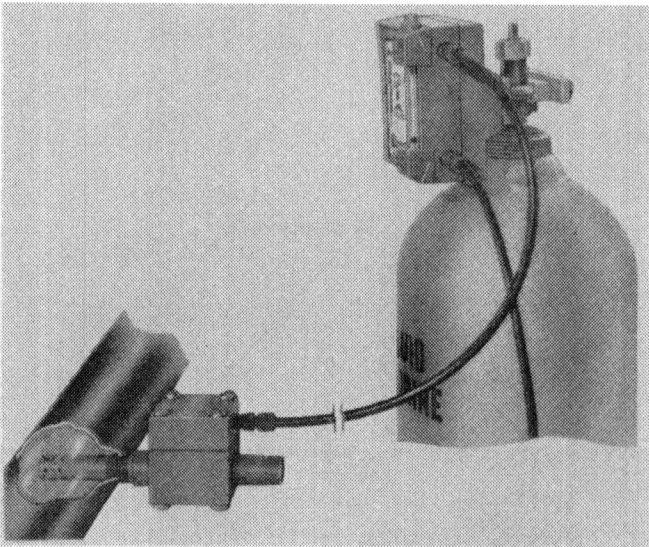

Courtesy of Capitol Controls Co.

FIG. 25.   VACUUM CHLORINATOR

The addition of ammonia to water prior to treatment with chlorine has been advocated by Babbitt et al. (1961), especially for supplies of water low in organic matter except when the detention period is brief or the pH is above about 8.0. These authors state that the procedure has the following advantages: (1) it decreases the development of off-flavors, especially those due to chlorinated phenols; (2) it improves the control of microorganisms in settling basins, filters, and the distribution systems due to the feasibility of dosing heavily with chlorine and the carrying of high chlorine residuals without the production of off-tastes; (3) it may give an improved bactericidal effect over the same amount of chlorine if considerable amounts of organic matter are present; (4) there tends to be a longer persistence of the bactericidal effect so that aftergrowths are inhibited; and (5) chlorine requirements are reduced. The ratio of chlorine to ammonia should not be less than 3 to 1. At pH 9.0 it may be 5 to 1 and at pH 5.0 it may be 9 to 1.

Chlorine has other functions besides disinfection. Its oxidant action is useful in removing iron and manganese and for converting the organic materials responsible for taste, odor, and color into imperceptible compounds (under certain conditions). The use of chlorine for removing taste and odor vectors demands careful control. Not only does chlorine have a taste and odor of its own which can be quite objectionable if the concentration becomes excessive, but the products of its reactions with organic compounds may have worse organoleptic properties than the original substances.

If progressive increments of chlorine are added to water containing organic contaminants, it is commonly observed that the off-flavors will first increase to a level higher than the initial and then decrease to a relatively low level.

It is often desirable to have a carryover of chlorine which will exert a continuing disinfectant action at subsequent points in the treatment, storage, and distribution system. This "residual" chlorine is obtained by supplying a dosage in excess of the immediate demand of the water and is of two types: (1) free residual chlorine; and (2) combined residual chlorine. Combined residual chlorine results when sufficient chlorine is added to combine with all of the ammonia which is present in most ground water and virtually all surface water. Free residual chlorine results when there is applied a dose of the chemical larger than the amount required for combination with the ammonia and organic matter. Both chlorine residuals have bactericidal and oxidative potential. The speed with which these reactions proceed depends upon the type of residual.

When the dosages of chlorine applied to a water containing ammonia and organic matter are plotted against the residual chlorine determined by the iodometric test or some other reliable method two inflections will be seen in the curve (see Fig. 26). A gradually rising combined available chlorine residual will be observed up to a certain maximum point, as expected. Beyond this point, further doses of chlorine result in a decline in the residual values. This is explained as being due to a more complete reaction with the ammonia or organic matter as a result of the higher total concentration of chlorine. The reactions involved have not been completely elucidated. The decline in residual chlorine continues to a second inflection point, the break point. After the minimum chlorine residual has been reached, further additions result in the development of a free available residual which can be determined chemically or electrometrically.

This residual increases in proportion to the subsequent additions of disinfectant. Oxidation of taste- and odor-causing substances takes place as the free available residual is developed. Since the oxidizable materials have been reacted with the maximum amount of chlorine, the residual is quite stable and remains available for inactivating any organisms which may enter the supply at some further point along the line.

There are several methods for determining residuals (Marks *et al.* 1951). Ghelberg (1964) described an amperometric titration method for free residual chlorine in water. The titration was carried out with 0.002256 *N* sodium arsenite. The current indicator was a cell containing the Au| Cl water, KCl, AgCl| Ag electrode system immersed in the water. The current was measured by a galvanometer of 10^{-7} sensitivity.

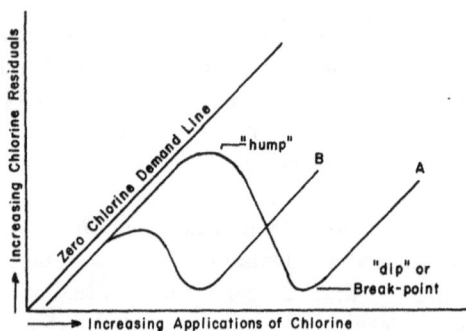

Fig. 26. The Relationship of Chlorine Dosage to Chlorine Residual in Two Samples of Water

The ammonia content of sample A is twice as great as that of sample B. As the ammonia content is increased, the amount of chlorine which must be applied to reach the initial peak is increased, and the height of the peak (the chlorine residual) is also increased.

From Griffin and Baker (1959)

Bacterial tests of the treated water are used to establish the residual chlorine level necessary to assure complete disinfection. A residual of 0.1 to 0.3 p.p.m. chlorine is usually adequate after ten minutes at 68°F. or higher. If the pH is greater than 7.0 or if the residual is in the form of combined available chlorine, higher levels are often necessary. The requirement is also markedly increased at lower temperatures. A satisfactory control technique for establishing the desirable chlorine residual in a given water supply is to determine the minimum necessary for bacterial kill and the maximum which can be tolerated from the standpoint of taste, and then apply chlorine in an amount sufficient to obtain a residual lying within this range. Economic considerations will dictate use of the lower end of the range with allowance of a workable margin of safety.

Chlorination of in-plant water supplies at concentrations in excess of the break-point is frequently practiced in the canning and freezing industries, based on suggestions of Hall and Blundell (1946). A continuous supply of germicidal chlorine is thus brought into contact with the equipment during its operation. Accumulation of microbial slimes is prevented or greatly reduced and bacterial counts of the finished products are reduced. Time required for cleanup is shortened. Many authors have cautioned that chlorine must not be used indiscriminately. Flavor, appearance, and, perhaps, some nutritional factors may be adversely affected by excessive dosages.

According to Somers (1951), in-plant chlorination of water during the preparation of canned foods had very little effect on the ascorbic acid content or the pH of the finished products when concentrations of up to 50 p.p.m. (the highest level studied) were used. There was an effect on flavor, however, and this was sometimes observed at concentrations as low as two parts per million as shown in Table 12.

TABLE 12

EFFECT OF CHLORINE TREATMENT ON FLAVOR OF CANNED FOODS[1]

| Product | Lowest Concentration Which Produced Off-flavor When 2, 5, 10, and 50 p.p.m. of Chlorine Were Added | |
	Partial treatment[2], P.p.m.	Complete treatment[3], P.p.m.
Applesauce, Rome Beauty[4]	10	5
Applesauce, Gravenstein[4]	(None at 50)	10
Apricots, halves unpeeled	"	50
Apricots, whole peeled	"	50
Asparagus, all green	50	50
Beans, green cut	50	10
Beans, green lima	50	10
Beans, with port, recanned[4]	. . .	50
Beets, red slices	50	10
Carrots, sliced	(None at 50)	10
Carrots, pureed[4]	"	50
Cherries, Royal Anne	"	50
Corn	. . .	(None with 15)
Figs, whole Kadota	50	5
Grapefruit juice, recanned[4]	. . .	50
Orange juice, recanned[4]	. . .	50
Peaches, clingstone halves	(None at 50)	5
Peaches, Elberta halves	"	10
Peas	. . .	(None with 10)
Pears	50	2 to 5
Pineapple juice, recanned[4]	. . .	10
Potatoes, sweet, solid pack[4]	(None at 50)	50
Pumpkin, solid pack[4]	"	50
Prunes, Italian	"	10
Spinach	50	10
Strawberries, whole	(None at 50)	5 to 10
Tomato juice[4]	. . .	10
Vegetable juice cocktail[4]	. . .	5
Yams, syrup pack	. . .	5

[1] From Somers (1951).
[2] Chlorination of all water except brines and syrups.
[3] Chlorination of all water, brines, and syrups.
[4] Chlorine was added directly to the product.

Chlorine dioxide is more expensive than chlorine and its application is more difficult to control, but it has some advantages which have led to its use for water treatment in special circumstances. Its oxidizing ability is high but it does not react with ammonia. It does, however, react with some nutrient factors, such as ascorbic acid, and this may be important in food applications. In contrast with chlorine, it is a good bactericide at high pH values. Chlorine dioxide oxidizes phenols without forming the highly odorous chlorophenols which can be perceived in the early stages of chlorination. If phenols are a serious problem in the water supply, free residual chlorination may be practiced to accomplish the usual objectives of this treatment, and then chlorine dioxide applied to oxidize the chlorophenols which have been formed.

The most common method for generating chlorine dioxide is the chlorine-chlorite process in which solutions of chlorine and of chlorite are mixed and put through a reaction tower. The reaction proceeds as follows:

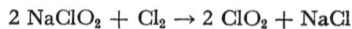

$$2 \ NaClO_2 + Cl_2 \rightarrow 2 \ ClO_2 + NaCl$$

The pH of the reaction mixture must be no higher than 4.0 if satisfactory yields are to be obtained. Use of equal parts of chlorine and sodium chlorite assures the development of a pH below this level (Feuss 1964).

Welch and Folinazzo (1959) made a study of the utility of chlorine dioxide in cannery sanitation and water conservation. They found that treatment of once-used water was highly effective in controlling bacteria and slime formation in pea and corn canneries. The persistent residuals and the lack of reaction with ammonia nitrogen made possible one-point rather than multiple-point application as required in rechlorination of used water. Total bacteria counts on the product were significantly lower when chlorine dioxide was used. However, chlorine dioxide was not considered more effective than chlorine in controlling thermophilic flat-sour spores. Chlorine dioxide residuals were more easily maintained in the presence of organic matter without generation of nitrogen trichloride so that, indirectly, chlorine dioxide might be considered a more effective sporicide.

POTASSIUM PERMANGANATE

Potassium permanganate as an agent for taste and odor control has been described in the preceding chapter. The substance also acts as a disinfectant, evidently as the result not only of its oxidizing action but also of the coagulating effect of the hydrous manganese floc formed when the permanganate is reduced. Because of its bactericidal action, permanganate has been suggested as a replacement for chlorine in certain applications.

Cleasby *et al.* (1964) made a thorough study of the permanganate kill of dilute cultures of *E. coli* at 68°F. Variables were: pH—5.9, 7.2, and 9.2; potassium permanganate concentrations—1, 2, 4, 8, 12, and 16 p.p.m.; and time—4, 8, 16, 30, 60, and 120 min. In the entire series, only a few combinations resulted in 100% kill. However, all tests at pH 5.9 showed at least a 99% reduction in *E. coli* counts at 120 min. When the temperature was increased, there was a slight but inconsequential improvement in kill.

These authors concluded that permanganate is not a good post-treatment disinfectant for two reasons: the brown stain produced by manganese dioxide prevents the use of a residual concentration in a distribution system, and the usual pH of treated water is not the most effective disinfecting pH for permanganate. The decline in potency with time under normal conditions of industrial use is greater than chlorine exhibits. The cost is also greater than that of chlorine for equivalent dosages.

FILTRATION

Microfiltration can be used to remove bacteria from water supplies but its action is relatively slow and many viruses are not held back. Berkfeld

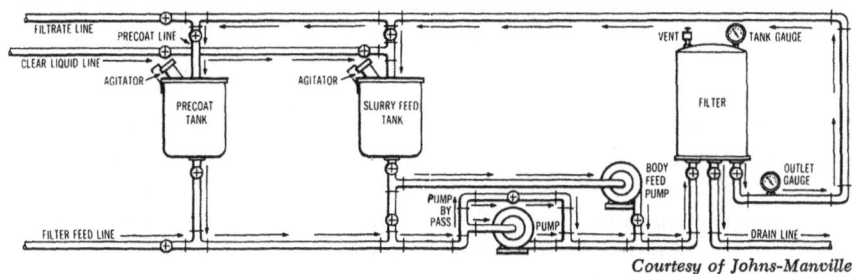

Courtesy of Johns-Manville

Fig. 27.   Method of Applying Diatomaceous Earth Filter Aids

and Pasteur filters have been known for many years as satisfactory devices for purifying relatively small volumes of water. In some cases they have been applied to commercial "sterilization" of fluids. The functional element in a Pasteur filter is a closed cylinder or thimble of unglazed porcelain. In the Berkfeld filter, it is a bed of fine infusorial earth. If properly constructed, these filters will deliver bacteria-free water when first put into service. After a time, the element becomes thoroughly saturated with microorganisms, the rate of flow is substantially reduced, and bacteria begin to appear in the filtrate. The apparatus can be re-activated by cleaning and sterilizing it or by applying temperatures high enough to volatilize the organic matter.

Filter membranes of plastic or metal with closely controlled pore size have become available in recent years. They are generally more reliable and give a higher flow rate than the Berkfeld or Pasteur devices. It does not appear that equipment of this type has been applied to high volume work in municipal or food-plant systems, and certainly not as a substitute for chlorination. Diatomite filter beds are used as an extra purification step in critical applications where water nearly devoid of microscopic particles is required, but only rarely in food processing plants. (See Fig. 27 for diagram of a system for applying diatomaceous filter aid.) Membrane filters have been used successfully as a substitute for the pasteurization step in beer processing.

The action of disinfecting filters is not merely a mechanical straining or sieving process. The bacteria appear to be adsorbed on the surface of the material lining the pores, probably through electrostatic attraction.

ALGAECIDES

Reservoirs and storage tanks in which water surfaces are exposed to sunlight may become fouled by heavy growths of algae. These organisms cause unpleasant odors, especially as they die and decompose. Free available residual chlorine of about 0.5 p.p.m. will usually prevent these growths if the concentration is maintained constantly throughout the reservoir. Sunlight tends to decompose chlorine and, as a practical matter, it may be rather difficult to maintain adequate levels of chlorine in all sections of large volumes of water.

Addition of copper sulfate has been found to be an effective method of controlling algae growth. The amount needed depends in part upon the type of organism, some being much more resistant than others to the toxic action of cupric ion, but a fraction of a part per million is often effective. In larger reservoirs, the chemical may be distributed in solid form from a boat. In smaller reservoirs, tanks, etc. the copper compound may be metered into the inlet as a concentrated solution.

BIBLIOGRAPHY

ADAMS, B. A. 1931. Substances producing taste in chlorinated water. Water and Water Eng. 33, 387–409.

BAKER, R. J. 1959. Types and significance of chlorine residuals. J. Am. Water Works Assoc. 51, 1185–1190.

BETZ, W. H., and BETZ, L. D. 1950. Betz Handbook of Industrial Water Conditioning. 3rd Edition. W. H. and L. D. Betz, Philadelphia.

BURTTSCHELL, R. H., ROSEN, A. A., MIDDLETON, F. M., and ETTINGER, M. B. 1959. Chlorine derivatives of phenol causing taste and odor. J. Am. Water Works Assoc. 51, 205–214.

CLEASBY, J. L., BAUMANN, E. R., and BLACK, C. D. 1964. Effectiveness of potassium permanganate for disinfection. J. Am. Water Works Assoc. 56, 456–474.

ETTINGER, M. B., and RUCHOFT, C. C. 1951. Effect of stepwise chlorination on taste- and odor-producing intensity of some phenolic compounds. J. Am. Water Works Assoc. *43*, 561–567.

FEUSS, J. V. 1964. Problems in determination of chlorine dioxide residuals. J. Am. Water Works Assoc. *56*, 607–615.

GHELBERG, N. W. 1964. The amperometric titration of free residual chlorine in water. Igiena *13*, No. 1, 63–66.

GRIFFIN, A. A. 1946. Break-point chlorination practices. Tech. Publ. No. *213*. Wallace and Tiernan, Newark, N. J.

GRIFFIN, A. E., and BAKER, J. R. 1959. The break-point process for free residual chlorination. J. New England Water Works Assoc. *73*, 1–17.

HALL, J. E. and BLUNDELL, C. C. 1946. The use of break-point chlorination and sterilized water in canning and freezing plants. Natl. Canners Assoc. Information Letter No. *1073* (February 12).

HARRIS, J. J. 1946. Chlorination in the food plant. Canner *103*, No. 9, 18–20.

INGOLS, R. S., and RIDENOUR, G. M. 1948. The elimination of phenolic tastes by chloro-oxidation. Water and Sewage Works *95*, 187–190.

INGOLS, R. S., WYCKOFF, H. A., KETHLEY, T. W., HODGDEN, H. W., FINCHER, E. L., HILDEBRAND, J. C., and MANDEL, J. E. 1953. Bactericidal studies of chlorine. Ind. Eng. Chem. *45*, 996–1000.

MARKS, H. C., WILLIAMS, D. B., and GLASGOW, G. U. 1951. Determination of residual chlorine compounds. J. Am. Water Works Assoc. *43*, 201–207.

MARKS, H. C., and STRANDSKOV, F. B. 1950. Halogens and their mode of action. Ann. N. Y. Acad. Sci. *53*, Art. 1, 163–171.

MERCER, W. A., and SOMERS, I. I. 1957. Chlorination in food plant sanitation. Advances in Food Research *7*, 129–169.

NESFIELD, V. 1962. The sterilization of drinking water by chlorine. J. Inst. Water Engrs. *16*, 217–222.

RITCHELL, E. C. 1947. Chlorination of canner water supply. Natl. Canners Assoc. Information Letter No. *1200*.

SOMERS, I. I. 1951. Studies on in-plant chlorination. Food Technol. 5, No. 2, 46–50.

WALTON, G. 1961. Effectiveness of water treatment processes as measured by coliform reduction. Public Health Service Publ. No. *898*.

WELCH, J. L., and FOLINAZZO, J. F. 1959. Use of chlorine dioxide for cannery sanitation and water conservation. Food Technol. *13*, 179–182.

CHAPTER 7

Recovery of Water from Processing Effluent and Sewage

INTRODUCTION

An often neglected source of water for food processing is waste liquid from antecedent operations in the same or adjacent plants. In a sense, most of the water obtained from surface sources is recirculated or diluted and stabilized sewage, since waste from disposal systems of communities and factories enters lakes and rivers from which other users draw their supplies.

In this chapter, sewage treatment is considered only insofar as it applies to recovery of water from contaminated effluent for re-use in food processing operations. Many procedures routinely used in treating sewage for discharge into streams, lakes, and other public water courses are also applicable to the renovation of waste from food plants.

The question may arise as to the need for considering methods of obtaining process water from processing effluents. It is true that the processor located in a region with ample supplies of water at reasonable cost and having a disposal outlet for all of his waste has no reason for using such methods. However, some raw water supplies are actually more contaminated than process effluent—and the number of these supplies is increasing constantly. Treatment of these effluents may be less costly than treatment of raw water. This is particularly true where the raw water is heavily contaminated with odorous organic substances or detergents, or is unusually high in salinity or hardness. Furthermore, the treatments required by local laws and regulations to be applied to process effluent before it can be dumped into streams may bring the liquid to a state of purity approaching that needed in the water supply. The cost of additional treatment which may be necessary can thus be brought within the acceptable range.

Industry consumes only about one-fiftieth of the water it withdraws. The rest is returned for downstream users. Clarke (1962) estimated that 80% of returned water has been used merely for cooling. It is "contaminated" only by heat, and the dissipation of heat through use of cooling towers and ponds is probably the most frequent example of water conservation by recycling. The other 20% of returned water carries a great variety of industrial wastes and other impurities. Some of these substances are relatively easy to remove by conventional techniques while others are extremely refractory and would require equipment of prohibitive cost for their extraction. Waste materials from food processing operations and those in ordinary domestic sewage are usually amenable to removal by the more common treatments even though quality may vary from the lightly contaminated cooling water to some of the thick sludges emanating from certain slaughterhouse departments.

Bulk handling methods used in food manufacture include use of water dumps for receiving raw fruits and vegetables and hydraulic conveyors for moving raw and partially processed materials to and from the various operations. It has been shown that microbial counts for foods in the final container are influenced by the condition of the water in which they are dumped and conveyed. The extent of the permissible recirculation in hydraulic conveying systems, and therefore the water demand of these systems, is dependent in part on the sanitary standards for the food product. For all practical purposes, the sanitary condition of the product can be measured in terms of the number of bacteria in or on the food discharged from the system (Geyer 1963).

Liquid wastes from canneries may vary from 0.25 to 4 gal. per lb. of fruits or vegetables processed. Water used in washing raw products may be high in inert solids (dirt) and contain insecticides or other chemicals. Biological methods of purification developed for treatment of domestic sewage are often inadequate for the stabilization of cannery wastes because of the poor balance of carbohydrates and nitrogenous compounds in the wastes (Hart and McGaukey 1964). The heavy seasonal loads imposed on municipal plants can easily disrupt treatment procedures. Combinations of screening, flotation, heavy chlorination, and chemical oxidants have been used to stabilize cannery effluent.

Disposal methods frequently used by food plants for liquid waste include lagooning, spray disposal, and digestion. In some cases, solids may be recovered to be sold as fertilizer or burned as fuel. Usually no thought is given to the recovery of the water present in these wastes.

Waste water, untreated or with primary treatment, may be suitable for irrigation. In the case of domestic sewage and the like, it is usually applied only to produce which does not go directly into the food supply, e.g., animal feed, pasture crops, and seed crops (Geyer 1963). In other countries, domestic sewage either untreated or treated by oxidation in pounds is routinely used for food crop irrigation (Hershkovitz and Feinmesser 1962).

If the wastes contain materials that may adversely affect plant growth, lagooning may have to precede irrigation. Molloy (1964) described a lagooning operation used as a primary treatment for wastes from instant coffee and tea manufacture. The settling, and presumably the oxidation, which occur in the lagoon over a period ranging from several hours to a few days, prepare the wastes for spray irrigation. The pH is 6.45, the solids content 0.59%, and the B.O.D. is 2,800 mg./l. in the liquid entering the lagoon, while the comparable figures for the spray liquid are 6.15, 0.43%, and 1,900 mg./l. The spray is not entirely innocuous after this simple treatment. It discolors, and in some cases kills, bushes and shrubbery although trees are not affected and grasses and other ground cover seem to thrive on it. The 1 to 3 in. of sludge which accumulates in the lagoon is removed by an annual cleaning.

Lagoons are ordinarily of little value for purification during the winter months in freezing climates. If the food processing is strictly seasonal this limitation may not be of particular importance, but in other cases it can be a serious difficulty. McIntosh and McGeorge (1964) describe the use of floating styrofoam rafts to cover the surface of a lagoon at Roby, Indiana (see Fig. 28). Before the installation of this insulation, B.O.D. was

From MacIntosh and McGeorge (1964)

FIG. 28. LAGOON PARTIALLY COVERED WITH INSULATING BLOCKS TO DECREASE EFFECT OF COLD ATMOSPHERE ON RATE OF DIGESTION

reduced only 32% in January operations. Reductions of about 68% were achieved after the modification.

Recovery or recycling of sewage has become practical and important to many industries. For example, the Bethlehem Steel Company plant in Baltimore uses about 125,000,000 gal./day of domestic sewage effluent for cooling and process needs. The receiving station obtains effluent from trickle filter and activated sludge treatment and applies supplementary settling and filtration before pumping it to the plant five miles away. Food plants are also using recycled or renovated effluent, although not, of course, in volumes as large as the quoted example.

Probably recycling is more common than renovation of waste water from outside sources. Clarke classifies on-site re-use systems as multiple recycle or cascade. Combinations of the two types are frequently used. A multiple recycle system consists of a number of parallel circuits carrying various grades of cooling waters and process waters. Such an arrangement segregates wastes for simplified treatment and provides for variation in temperature, pressure, and make-up water quality. One or more of the circuits may have integral purification stages for maintaining the necessary water characteristics. Usually, circuits are interconnected and have continuous blowdown drainage to limit retention time in a circuit and deliver the wastes to less critical systems or to waste treatment facilities. If the blowdown waste passes to another process or cooling system, the plant is a compound multicycle cascade type.

Cascade re-use systems depend on in-line integration of the water supply for two or more operators rather than differentiation of process and cooling waters. Water first enters the equipment or process requiring the cleanest or coldest supply and continues to successive operations where progressively higher temperatures and poorer water qualities can be tolerated. The effluent of one operation becomes the supply for the next until unacceptable temperature or contamination develops. In tandem re-use, one processor passes his effluent to another plant requiring a less pure supply.

Ground water recharge is a method of maintaining the water table or at least of reducing the rate at which the table is being lowered or invaded by saline intrusion in areas of heavy withdrawal. In essence, this procedure takes advantage of natural purification processes for recycling water. The water is retained in the general area of withdrawal so that it can contribute to the total reserve available to users in the area, whereas dumping the waste into a stream generally means the water is lost to that particular region. Obviously, ground water recharge is not a simple matter of pumping the raw sewage into a hole drilled into the earth. A system

for distributing the effluent over a sufficient area to assure gradual percolation to the water table so that the material is not forced to the surface,
proper treatment methods so that the natural purification steps can proceed rapidly, and isolation of withdrawal wells from the disposal area are
the minimum necessary precautions.

Occasionally, some particularly objectionable waste substance which in-
terferes with even limited re-use of the water can be recovered for sale or
use. The recovery of tartrates from winery wastes is an example of a
treatment step which may either pay for itself or yield a small profit.

The 1959 Census of Manufacturing Establishments shows that among
all food and allied producers using more than 20,000,000 gal./yr., 70% of
the firms were re-users of some part of process effluent.

METHODS OF TREATMENT

There is a continuing effort among water technologists to develop new
methods of waste treatment which will produce effluents of superior quality. However, the initial steps in treatment are still those which have been
used for years in municipal sewage plants. The two sewage treatments
which can be used are the trickling filter process and the activated sludge
system (Weiner 1963). In both of these procedures, waste is first passed
through a screen of metal bars that removes all solid objects too large to be
handled in subsequent operations. Usually, the second step is the passing
of the liquid through a long, narrow tank called a grit chamber where
large suspended particles settle out. If the content of such material is
very low, as in some reliably homogeneous processing wastes, the grit
chamber can be omitted.

The primary settling tank is a necessary part of the trickling filter process and is also used in most of the activated sludge systems. This is a
large, usually open, basin in which the liquid waste is held until the larger
suspended particles settle out and can be drawn off at the bottom as
sludge. These concentrated impurities are sent to a digester which is a
large cylindrical, sealed tank where they may be kept for about 30 days.
Under the conditions of low oxygen tension existing in the digester, anaerobic bacteria metabolize most of the soluble materials, hydrolyze some
of the insoluble matter, and convert part of the solids to gases such as
methane and hydrogen sulfide. These actions change the sludge into a
material of much greater biochemical stability.

The precipitated material remaining in the digester is dried and dis-
posed of in land fill, by burning, or as fertilizer, while the effluent is
drawn off and sent to the trickling filter.

The equipment which gives the trickling filter process its name is a large
bed of rocks, crushed stone, and gravel several feet thick. The discharge

from the digester is distributed over the top of the filter bed by sprinklers which can be of several designs. The gravel or other material in the filter becomes coated with a film of aerobic bacteria which act upon the metabolizable soluble compounds in the liquid. These compounds are converted into carbon dioxide, water, and cell components, as well as a few rather minor metabolic by-products. Since the cell components remain in the organisms attached to the gravel, they are effectively removed from the effluent which is then sufficiently pure to be discharged to a river or lake.

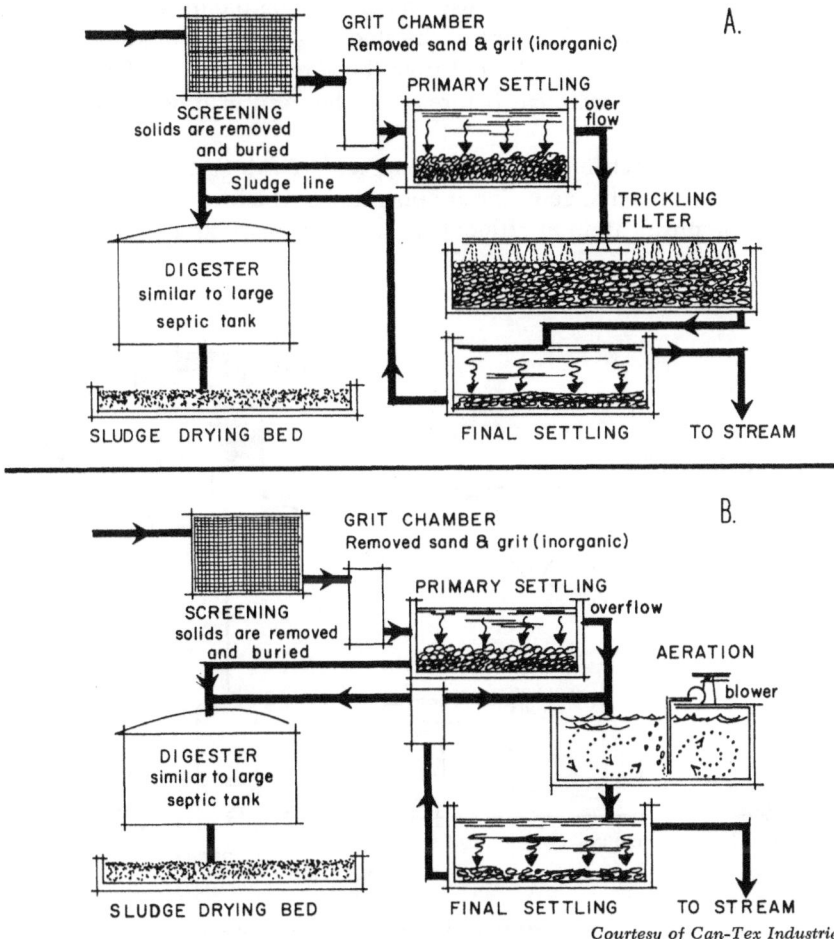

Courtesy of Can-Tex Industries

FIG. 29. TWO METHODS OF TREATING SEWAGE—THE TRICKLING FILTER PROCESS AND
THE ACTIVATED SLUDGE SYSTEM

One of the advantages of the activated sludge system (see Fig. 29) is that the amount of sludge going to the digester is considerably reduced so that the size of the digesters can be less than in the trickling filter method. The raw sewage or plant waste may, in the former process, be sent to a primary settling tank. This step may be omitted under certain conditions. In any case, the effluent next flows through a series of aeration tanks. Here it is agitated and oxygenated by air blown in from the sides and bottoms of the tanks. The conditions are extremely favorable for aerobic bacteria and they digest the wastes in a few hours. Sludge is removed in a settling tank and a small proportion of it—the activated sludge—is returned to the aeration tank where it assists in maintaining the microbiological activity at a high level. The rest of the sludge is sent to a digester for anaerobic stabilization. The effluent from the aeration tank is relatively pure but needs further treatment before it can be used in processing applications.

Coagulation, settling, chlorination, and treatment with activated charcoal or ion-exchange resins are operations which can be used in completing the purification of effluents from the primary and secondary sew-

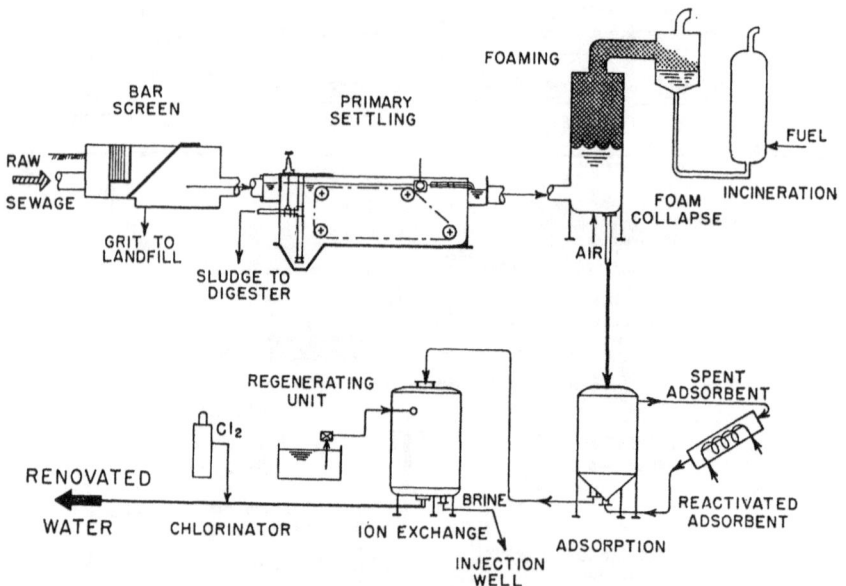

FIG. 30. POSSIBLE WATER RENOVATION PLANT

According to the designer (McCallum 1963), the diagram is meant to illustrate a number of separation and disposal processes and is not necessarily a suggested system.

age treatment processes described above. These do not differ in principle from the techniques described in Chapter 5. A possible water renovation plant using conventional and other techniques is shown in Fig. 30.

Culp (1963) described tertiary treatments for reclamation of waste water. The methods described, or some modification of them, have been tested in more than 350 runs at pilot plants at Corvallis and Salem, Oregon, with combined domestic and cannery wastes as influent, and at the South Tahoe Public Utility District in California and at Philomath, Oregon, treating domestic wastes only. The tests were directed toward removal of alkyl benzene sulfonate, reduction of phosphate concentration, and general over-all improvement of treated wastes. The tests used primary as well as secondary effluent from both trickling filter and activated sludge plants.

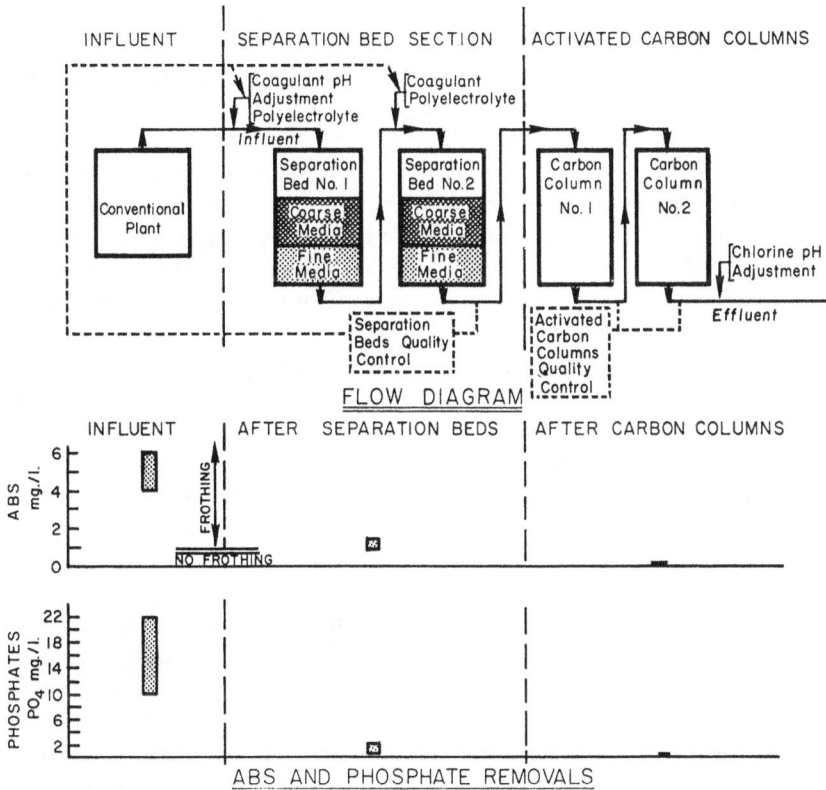

From Culp (1963)

FIG. 31. TERTIARY TREATMENT FOR WATER RECLAMATION

The supplemental treatment advocated by Culp for influents encountered and under conditions existing in the test plants included chemical coagulation with 100 to 200 mg./l. of alum and a coagulant aid, primary adsorption on alum floc, and rapid filtration followed when necessary by secondary adsorption on activated carbon. The improvements over conventional measures were concerned with adaptations of the individual processes to the specific problems involved in processing of effluent from primary and secondary treatment of sewage. Part of these adaptations consisted in newly designed equipment and an improved flow pattern (see Fig. 31).

The separation bed which removes the precipitate and associated suspended material after addition of the flocculants is similar in outward appearance to a rapid sand filter, but the bed is arranged so that the waste first contacts the coarse material and then is passed through layers of increasingly finer granulation. Filtering in the direction of coarse to fine reduces surface plugging which would otherwise occur and makes it possible to use the entire depth of the bed for removal. Loss of head proceeds at a slower rate for this reason. The size of the filtration media is selected so that the raw waste will pass through the bed with no loss of particles until the flocculants are added.

The separation beds were washed similarly to rapid sand filters, with surface washing a necessary step. Washing consumed about four per cent of the treated water. Regrading of the media after backwashing was simplified by using coarse and fine material of different specific gravities.

Effluent from the separation beds operated in the preceding manner exceeded in clarity many drinking water supplies, with typical turbidity values being less than 0.5 standard units. Phosphates were greatly reduced (average values about 1.0 mg./l. in the effluent) as a result of adsorption on the alum floc. Some of the alkyl benzene sulfonates were also removed and the B.O.D. values were reduced about 50%. These results are better than those to be expected from chemical precipitation, settling, and rapid sand filtration. Part of this superiority is attributed to the very favorable conditions for adsorption in the separation bed.

The carbon filtration section of the plants consisted of a series of two or more columns about three feet in depth filled with granular activated carbon and passing fluid at five gallons per square foot per minute. The carbon filtration efficiency and economy were greatly enhanced by a filtration bed which removed materials that would otherwise quickly foul the carbon particles. The final effluent contained nothing that would identify it as a waste. The alkyl benzene sulfonate concentration was below 0.05 mg./l., the phosphate content less than 0.5 mg./l., and the B.O.D. less than ten milligrams per liter. Chlorination could have been applied to render the water bacteriologically pure. There is little doubt that the chlorinated effluent from the system described by Culp would be suitable for many process applications in the food industry.

The pH of the system appears to have a strong influence on the effectiveness of the different treatment steps. Lea *et al.* (1954) indicated that phosphate adsorption on alum floc is optimal at pH 7.1 to 7.7.

LIMITATIONS OF RECYCLING

Continual recycling of wastes for a long time may lead to the gradual accumulation of contaminants which enter the system at a seemingly insignificant rate but which are removed incompletely, if at all, by the methods of purification. In the natural scheme of things, these contaminants would ultimately be flushed into the sea or decomposed by complex reactions occurring in the soil and streams.

A most interesting and informative study of difficulties involved in recycling sewage for an extended period with relatively minor contributions by natural or uncontrolled processes was reported by Metzler *et al.* (1958). Due to a prolonged drought, the normal water supply of Chanute, Kansas (river water) was almost completely abolished for nearly a year. By using the regular treatment facilities of the town plus various supplementary measures, the sewage was recycled for about five months and provided the principal source of potable water for the inhabitants during this time. The investigators observed a rapid increase in the concentrates of dissolved salts and organic materials, many of which were not amenable to removal by ordinary process. Although the organoleptic quality of water deteriorated markedly during the treatment period, the bacteriological quality as evaluated by the coliform test remained excellent. At the end of the recycling period, the treated water had a pale yellow color and an unpleasant musty taste and odor. It foamed when agitated.

Some of the difficulties observed by Metzler *et al.* during the prolonged recycling of domestic sewage at Chanute were: (1) there was an inability to practice free residual chlorination for taste and odor control and color removal because of the extremely high chlorine demand of the raw water; (2) frothing occurred at the recarbonation basin and in the rapid sand filters during backwashing; (3) coagulation and settling was less effective; (4) the filter sand became plugged and coated; and (5) false high readings of free-chlorine residuals were obtained by the OTA test.

Some industrial contaminants evidently persist unchanged through all treatment processes and for long periods in natural water courses. These may lead to undesirable odor, flavor, or appearance effects in commercial food and beverage products. They may also disrupt sewage treatment processes by killing off the bacteria responsible for oxidation of other waste compounds or by interfering with coagulation and sedimentation.

As pointed out by McCallum (1963), concentrated wastes removed by advanced waste treatment processes cannot be discarded into usable surface or ground supplies. Four approaches can be considered: (1) recovery of part of the concentrated waste for use as fertilizer, chemical raw materials, etc.; (2) disposal of the highly concentrated wastes into controlled waste sinks or into remote ocean areas; (3) subsurface disposal through injection into deep wells or sealed underground cavities; and (4) conversion of the contaminants to innocuous forms by digestion, incineration, or wet oxidation.

SOURCES OF EQUIPMENT

Treatment schemes of almost any desired degree of complexity can be implemented by packaged plants supplied by experienced firms in the sewage disposal field; or plants designed by consultants can be constructed from the basic units of equipment.

The Hydrodynamics Division of FMC Corporation manufactures package plants providing rapid and complete liquid sewage treatment and aerobic digestion of solids. The design is based in part on demonstration plant studies of factors affecting activated sludge processes which were conducted by engineers of the Société Génerale d'Épuration et d'Assainissement of Paris. Plants handling from 30,000 to 5,000,000 gal./day have been constructed.

Contac-Pac units of the Link-Belt Corporation are intended to accept from 5,000 to 15,000 gal./day in the one-tank style and from 15,000 to 30,000 gal./day in the two-tank version. Larger flows can be accommodated by using multiples of the two-tank system. The tanks are 11 ft. 6 in., or less, in diameter. The Contac-Pac process involves four separate stages: primary settling of solids, contact aeration of primary effluent, final settling of solids, and aerobic digestion of all settled solids (sludge). The four operations can be combined in a single tank. One of the distinctive features of the process is use of asbestos cement contact plates to facilitate growth and improve the activity of the microorganisms which oxidize the sewage. According to the manufacturer, this feature has the desirable treatment qualities of the activated sludge method while at the same time providing the settling characteristics of solids from a trickling filter.

BIBLIOGRAPHY

ACKERMAN, E. A., and LÖF, G. O. G. 1959. Technology in American Water Development. Resources for the Future, Inc., Washington.

ALLEN, L. A., BLEZARD, N., and WHEATLAND, A. B. 1946. Toxicity to fish of chlorinated sewage effluent. Surveyor 105, 298–300.

ANON. 1955. Standard Methods for the Examination of Water, Sewage, and Industrial Wastes. Tenth Edition. American Public Health Association, Inc., New York.

ANON. 1963. Re-use of Water in Industry. Butterworths, London.

BABBITT, H. E., and BAUMANN, E. R. 1958. Sewerage and Sewage Treatment. Eighth Edition. John Wiley and Sons, New York.

BAITY, H. G., MERRYFIELD, F., and UZZLE, A. B. 1933. Some effects of sewage chlorination upon the receiving stream. Sewage Works J. 5, 429–440.

BARDUHN, A. J., ROSE, A., and SWEENEY, R. F. 1963. Waste-water renovation. Public Health Serv. Environmental Health Ser. AWTR-4.

CLARKE, F. E. 1962. Industrial re-use of water. Ind. Eng. Chem. 54, No. 2, 18–27.

CONLEY, W. R., and PITMAN, R. W. 1960. Innovations in water clarification. J. Am. Water Works Assoc. 52, 1319–1330.

CULP, R. L. 1963. Wastewater reclamation by tertiary treatment. J. Water Pollution Control Fed. 35, 799–806.

DEUTSCH, M. 1962. Controlled induced-recharge tests at Kalamazoo, Michigan. J. Am. Water Works Assoc. 54, 181–190.

DIXEY, F. 1931. A Practical Handbook of Water Supply. Thomas Murly and Co., London.

FAIR, G. M., and GEYER, J. C. 1954. Water Supply and Waste Water Disposal. John Wiley and Sons, New York.

FAIR, G. M., and GEYER, J. C. 1958. Elements of Water Supply and Waste Water Disposal. John Wiley and Sons, New York.

GAINEY, P. L. 1950. An Introduction to the Microbiology of Water and Sewage for Engineering Students. Burgess Publishing Co., Minneapolis.

GERSTER, J. A. 1963. Cost of purifying municipal wastes by distillation. U.S. Public Health Serv. Environmental Health Ser. AWTR-6.

GEYER, J. C. 1963. Water conservation. National Research Council Publ. 942, 17–49.

GURNHAM, C. F. 1955. Principles of Industrial Waste Treatment. John Wiley and Sons, New York.

HART, S. A., and McGAUHEY, P. H. 1964. The management of wastes in the food producing and food processing industries. Food Technol. 18, 432–438.

HERSHKOVITZ, S. Z., and FEINMESSER, A. 1962. Sewage reclaimed for irrigation in Israel farm oxidation ponds. Wastes Eng. 33, 405–410.

HINDIN, E., MAY, D. S., McDONALD, R., and DUNSTAN, G. H. 1964. Analysis of volatile fatty acids in sewage by gas chromatography. Water and Sewage Works 111, 92–95.

HYNES, H. B. N. 1960. The Biology of Polluted Waters. Liverpool Univ. Press, Liverpool, England.

JOHNSON, R. L., LOWES,, F. J., Jr., SMITH, R. M., and POWERS, T. J. 1964. Evaluation of the use of activated carbons and chemical regenerants in treatment of waste water. U. S. Public Health Serv. Environmental Health Ser. AWTR-11.

JOYCE, R. S., and SUKENIK, V. A. 1962. Activated carbon in waste-water treatment—adsorption and thermal reactivation. Seminar on Advanced Waste Treatment Research, Cincinnati, Ohio, May 15–16, 1962.

JOYCE, R. S., and SUKENIK, V. A. 1964. Feasibility of granular, activated-carbon adsorption for waste-water renovation. U.S. Public Health Serv. Environmental Health Ser. AWTR-10.

KEEFER, C. F. 1962. Tertiary sewage treatment. Public Works 93, No. 11, 109–120; No. 12, 81–90.

KOENIG, L. 1964. Ultimate disposal of advanced treatment waste. Public Health Serv. Environmental Health Ser. AWTR-8.

KUENEN, P. H. 1955. Realms of Water. John Wiley and Sons, New York.

LEA, W. L., ROHLICH, G.A., and KATZ, W. J. 1954. Removal of phosphates from treated sewage. Sewage and Industrial Wastes 26, 261–270.

LEGGAT, R. 1949. The chlorination of sewage effluents. J. Research Sanit. Inst. 69, 11–20.

McCABE, J., and ECKENFELDER, W. 1958. Biological Treatment of Sewage and Industrial Waste. Reinhold Publishing Corp., New York.

McCALLUM, G. E. 1963. Advanced waste treatment and water re-use. J. Water Pollution Control Federation 35, 1–10.

McINTOSH, G. H., and McGEORGE, G. G. 1964. Year round lagoon operation. Food Processing 25, No. 1, 82–86.

McRAE, W. A. 1962. Liquid-phase permeation through membranes. National Research Council Publ. 942, 225–269.

MERCER, W. A. 1964. Physical characteristics of recirculated water as related to their sanitary condition. Food Technol. 18, 335–340, 344.

MERCER, W. A., CHAPMAN, J. E., ROSE, W. W., KATSUYAMA, A., and DWINNELL, F., JR. 1962. Aerobic composting of vegetable and fruit wastes. Compost Sci. 3, 9–20.

METZLER, D. F., CULP, R. L., STOLTENBERG, H. A., WOODWARD, R. L., WALTON, G., CHANG, S. L., CLARKE, N. A., PALMER, C. M., and MIDDLETON, F. M. 1958. Emergency use of reclaimed water for potable supply at Chanute, Kans. J. Am. Water Works Assoc. 50, 1021–1060.

MOLLOY, D. J. 1964. "Instant" wastes treatment. Water Works and Wastes Eng. 1, No. 2, 68–70.

MORRIS, J. C., and WEBER, W. J., JR. 1962. Preliminary appraisal of advanced waste treatment process. U.S. Public Health Serv. Advanced Waste Treatment Research Publ. 2.

NEALE, J. H. 1964. Advanced waste treatment by distillation. U.S. Public Health Serv. Environmental Health Ser. AWTR-7.

OSWALD, W. J., GOLUEKE, C. G., COOPER, R. C., and BRONSON, J. C. 1962. Water reclamation, algal production, and methane fermentation in waste ponds. J. Water Pollution Control Fed. 34, 234–240.

REID, G. W. 1949. Adsorption and assimilation of P[32] by bacterial slimes. Tech. Information Div., Oak Ridge Laboratories, Oak Ridge, Tenn.

RUBIN, E., EVERETT, R., JR., WEINSTOCK, J. R., and SCHOEN, H. M. 1963. Contaminant removal from sewage plant effluents by foaming. Public Health Serv. Environmental Health Ser. AWTR-5.

VARMA, M. M., and REID, G. W. 1964. Comparison of respiration and metabolism of biological slimes using radiophosphorus. J. Water Pollution Control Fed. 36, 176–200.

WALTON, G., and CULP, G. L. 1964. Chlorine disinfection in primary sewage treatment—a review of the literature. Water and Sewage Works 111, 80–81.

WEINER, D. L. 1963. Understanding the World of Sewerage. Can-Tex Industries, Inc., Cannelton, Ind.

WELCH, J. L. 1959. Use of chlorine dioxide for cannery sanitation and water conservation. Food Technol. *13*, 179–182.

WILLIAMSON, J. N., HEIT, A. H., and CALMON, C. 1964. Evaluation of various adsorbents and coagulants for waste-water renovation. U.S. Public Health Serv. Environmental Health Ser. AWTR-12.

CHAPTER 8

Ingredient (Potable) Water from
Saline Sources

INTRODUCTION

A factor limiting economic growth—including expansion of food processing facilities—in many regions of the world is the local supply of potable water. Furthermore, several studies published in recent years have purported to show that some areas in the United States face the prospect of a shortage of potable water in the near future as a result either of the depletion of present supplies or of unavoidable increases in consumption. Although the total consumptive demand for water in the United States as a whole is only about 10 to 20% of the total fresh water supply economically available at the present time, ample supplies are not always present where they are needed. The situation is aggravated in many cases by pollution or by intrusion of sea water due to lowering of the water table.

Often these problems could be alleviated if the oceans (or other saline sources, such as brackish wells, springs, and lakes) could be drawn upon as sources of potable water. However, the salinity of untreated sea water make it useless not only for drinking but also for nearly all industrial uses except (under some conditions) cooling. Desalination of such water involves procedures rarely applied to water from the usual sources, and application of these processes necessitates heavy expenditures of energy and the construction of elaborate and costly plants. Other treatments which must be applied to sea water to make it potable (e.g., the removal of color and odor or the destruction of microorganisms) can be accomplished through the use of more or less conventional methods at acceptable costs.

There is no expectation that potable water from the sea or other saline source will ever be able to compete in price with that from ground or surface supplies where ample quantitites of the latter are available. According to a recent survey by the American Water Works Association of 497 utilities in the United States (Anon. 1961), the average cost of 1,000 gal. of water to public suppliers was 12.3 cents, including not only the costs of treatment but also those of distribution, depreciation of plant equipment, and administration. The cost of treatment for fresh water, including softening, approximates five cents per 1,000 gal.

It has been calculated that the absolute minimum energy requirement for separating 1,000 gal. of fresh water from a reservoir of sea water by an ideal process with no losses or inefficiencies would be 2.5 kw. h. (Rothbaum 1958). The minimum energy requirement for any conceivable practical method would be perhaps four times as great, say about 10 to 12 kw. h. Dodge (1962) estimated that only about 10% thermodynamic efficiency can be attained with processes that we now have. It is quite apparent that the economic success of any commercial desalination plant would depend largely on the relative value of available energy and potable water in any given area. According to Chirico (1962), there are good reasons for believing that the cost of conversion can be brought to as low as 35 cents per 1,000 gal. when dual purpose (electric power and desalination) plants in the capacity range of 15 to 20 million gal. per day have been built.

Ackerman (1963) calculated that conversion of saline aquifers would be widely applicable now or in the very near future to small industrial water supplies in certain areas in the northern Great Plains–Great Basin area if costs between 50 and 70 cents per 1,000 gal. could be achieved. Similarly, Koenig (1959) indicated that saline water could be economical in comparison to conventional supplies for system capacities of 200,000 to 300,000 g.p.d. where no surface supply was available within 1 to 10 miles. The cost of converting sea water was reduced from about $5/1,000 gal. in 1950 to $1 per 1,000 gal. in 1961 with a potential of perhaps 60 cents/1,000 gal. in 1965–1967 (Gilliam 1963), but the latter figure is not likely to be significantly reduced with single-purpose plants for many years, if at all.

Desalting plants have been in operation for several years in the Virgin Islands, in Kuwait and Bahrein in the Persian Gulf area, in Aruba of the West Indies, in Tobruk (Libya), in Welkom (Union of South Africa), and Eilat (southern Israel) according to Spiegler (1962). Sea-water distillation aboard ocean-going vessels has been standard practice for over a century. These are all relatively high cost installations supplying water so expensive that it would be acceptable in very few places in the continental United States. The possibility that low-cost saline water conversion would be of great benefit to many regions of this country has led to government-sponsored studies of possible means for cheapening desalination to bring it within an acceptable price range. Considerable success has been achieved in cost reduction in demonstration plants constructed by the Department of Interior's Office of Saline Water.

It has been suggested by Horning (Anon. 1964) that very large atomic power plants coupled with water desalting plants could become economically feasible about 1975 to 1980. This suggestion envisions a combination atomic power-plant and desalination unit having an output of 500 million gallons of fresh water per day at a cost of 20 to 25 cents per 1,000 gal.

Water is not considered potable when the concentration of salts exceeds 1,000 p.p.m. and much lower concentrations are desirable for many food

processing operations.   Ocean water contains about 3.5% salts of which
approximately 75% is present as sodium and chloride ions.   The ten most
abundant elements present in solution in sea water are given in Table 13.

TABLE 13

CONCENTRATIONS OF THE TEN MOST ABUNDANT ELEMENTS IN SEA WATER[1]

Element	Concentration Mg./Kg.[2]
Chlorine	18,980
Sodium	10,561
Magnesium	1,272
Sulfur	884
Calcium	400
Potassium	380
Bromine	65
Carbon	28
Strontium	13
Boron	4.6

[1] Sverdrup *et al.* (1942).
[2] Calculated on the basis of a chlorinity of 19.00.

The amounts of dissolved gases are not included in these figures.   Relative proportions of the main constituents are about the same in all oceans
although some variation in total solids has been found in different regions.

Conversion of saline water can be experimentally approached from
either of two directions: removal of salts from water or the removal of
more or less pure water from the raw liquid.   When water molecules are to
be acted upon, unit operations of vaporization, crystallization (freezing or
hydrate formation), solvent extraction, reverse osmosis, or absorption can
be utilized.   When salt ions are to be transferred, ion exchange or semipermeable membrane techniques such as electro-dialysis may be applied.
Biological desalting and other exotic processes have also been investigated
in preliminary studies without practical applications having been made.
Distillation and electrodialysis are already in use in commercial installations and are by far the most technically advanced methods available at
the present time.

Water can be desalted by insolubilizing the ions through chemical reactions.   Using such a principle, the Armed Forces have developed desalting
kits for use in life rafts and in other emergency situations where sea water
is the only water available for drinking.   The kit contains a briquet having
as its main active ingredient a silver zeolite which reacts with the chlorides
of sodium, magnesium, and calcium to form insoluble silver chloride and
insoluble sodium, magnesium, and calcium zeolite.   The probable reactions can be summarized by the following equations, in which Z indicates
a zeolite:

TABLE 14

ENERGY REQUIREMENTS AND INVESTMENT COSTS OF SELECTED DESALTING PLANTS[1]

Method	Capacity,[2] Cu. M. P. D.	Energy Requirements		Efficiency,[3] %	Investment, Dollars Per Daily Cu. M.	Remarks
		Kw. H./ Cu. M.	Tons Steam Per Cu. M.			
Six-effect still	12,000	...	0.21	3.1	1,000	Three parallel units in Aruba, Netherlands Antilles
Twelve-effect still	4,000	...	0.09	7.2	330	Demonstration plant. Investment includes special features
Multistage flash	4,000	...	0.17	3.8	410	Estimate from results with smaller existing plants
52-stage flash	185,000	...	0.073	8.9	170	Design study for plant coupled to nuclear reactor
Compression still	4,000	21	...	4.6	340	Forced circulation vapor compression
Rotary compression still	250	17[4]	...	5.7	460	Estimates based on results with smaller units
Electrodialysis (sea water)	10–30	36	...	2.7	1,100	Small plants
Electrodialysis of brackish water, 2800–800 p.p.m.	11,000[5]	3.15	...	2.4	73	Nine water treatment plants in South Africa
Indirect freezing	40,000	6.9	...	14.	270	Estimated from pilot plant studies
Solar evaporation (glass-covered stills)	...	...	...	1	7,500	Considerably lower investment for plastic equipment

[1] Adapted from Spiegler (1962). See this reference for sources of original data.

[2] One cu. m. equals 264.17 U. S. gallons. All volumes refer to the fresh water product.

[3] The efficiency is the ratio, expressed as percentage, of (a) the theoretical minimum energy requirement for splitting salt water of 77°F. into fresh water and a brine of doubled concentration and (b) the actual energy requirement including pumping within the plant. Assumptions are: (1) Half of the water is discharged as fresh product and the other half as brine, except for the dialysis of brackish water which is concentrated four times; (2) Theoretical power requirement equals 0.97 kw. hr./cu. m. except for brackish water which theoretically requires 0.076 kw. h./cu. m.; (3) Each latent calorie of steam is equivalent to one-quarter calorie of electric power, e.g., at 100% efficiency each ton of steam can produce 154 tons of fresh water.

[4] Experimental figure obtained with a plant of only 60 cu. m./day capacity. Lower power consumption for a 250 cu. m./day plant is possible. Inventor's estimate for larger plants is 13 kw. hr./cu. m.

[5] Designed capacity; in practice, only 55% utilized.

$$2NaCl + Ag_2Z \rightarrow NaZ + 2AgCl \qquad (1)$$

$$MgCl_2 + Ag_2Z \rightarrow MgZ + 2AgCl \qquad (2)$$

$$CaCl_2 + Ag_2Z \rightarrow CaZ + 2AgCl \qquad (3)$$

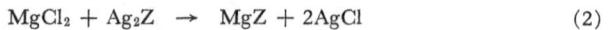

A small quantity of barium hydroxide may be incorporated in the briquet to reduce the sulfate content. The barium ions react with sulfate ions to form insoluble barium sulfate while Mg^{++} reacts with hydroxide to form insoluble $Mg(OH)_2$. In addition, small amounts of activated carbon for flavor improvement and a disrupter which helps to disintegrate the briquet when it comes into contact with water are included. It is quite apparent that desalination by this method is very expensive and justifiable only under conditions of extreme emergency.

DISTILLATION

Vaporization-condensation, or distillation, is an obvious approach to separating pure water from a reservoir of brine. Since sea water and most brackish supplies contain only traces of other condensable materials which are volatile at the boiling point of water, a very pure product can be obtained in one step by means of distillation procedures. The practical difficulties of engineering an economically acceptable plant are considerable, however. Development work on distillation processes has been directed mostly to designing apparatus which is energetically more efficient and to overcoming the problem of scale formation. Considerable success has been achieved in these areas.

Dodge (1962) lists seven major variants of stills and distillation processes: (1) multiple-effect, boiling type; (2) multiple-effect, flash type; (3) vapor compression; (4) rotating or wiped surface; (5) indirect transfer of heat, both sensible and latent, by an insoluble liquid; (6) direct transfer of heat from an evaporating to a condensing surface; and (7) solar energy processes.

In multiple-effect distillation processes of the boiling type (Fig. 32), the vapor produced by heating water in one evaporator chamber in a series is used as the heat source for the following evaporator. The pressure in the latter is lowered so that the brine will boil, using the latent heat of condensation of vapor as its heat supply.

In the flash-type of multiple-effect distillation (Fig. 33), heated saline water is introduced into a chamber maintained at reduced pressure. These conditions cause an immediate flashing or boiling of part of the water. The remaining brine is passed through a series of chambers at progressively reduced pressures. In each chamber, an additional portion of water is converted into steam. Heating of the sea water is accomplished

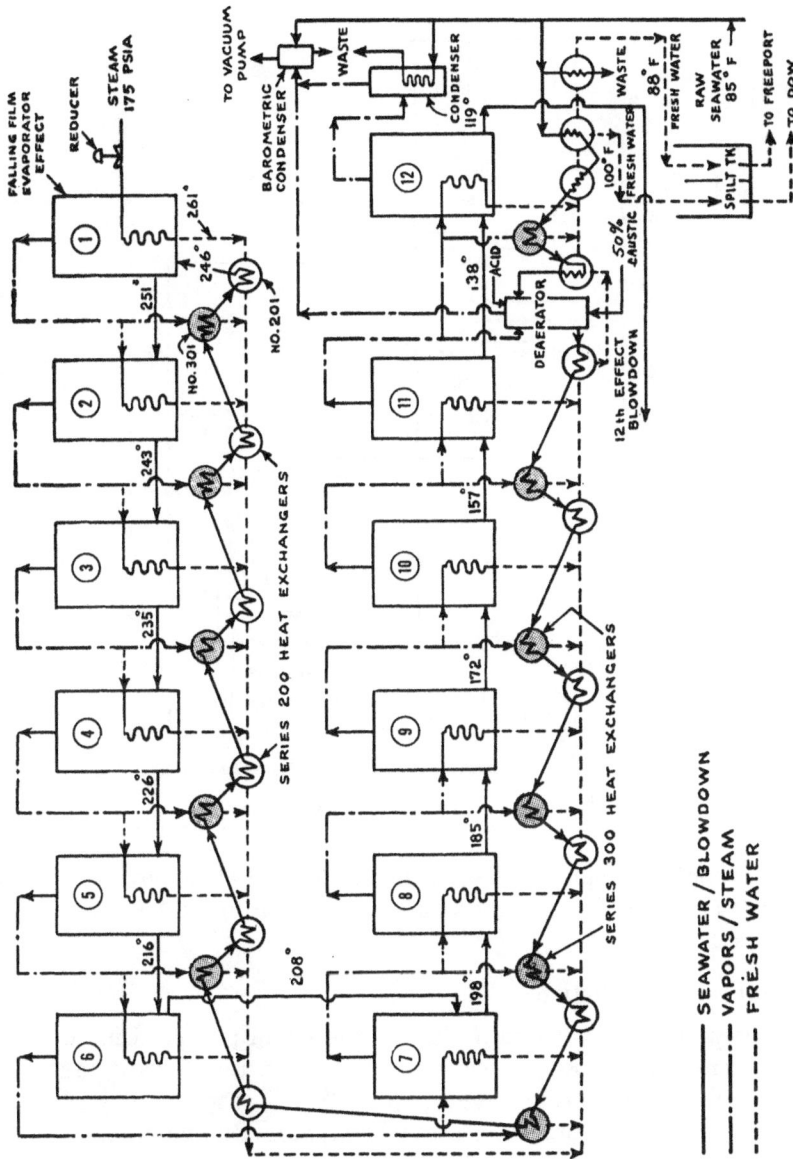

FIG. 32. SIMPLIFIED FLOWSHEET OF A MULTIPLE-EFFECT EVAPORATION PLANT OPERATING AT FREEPORT, TEX.

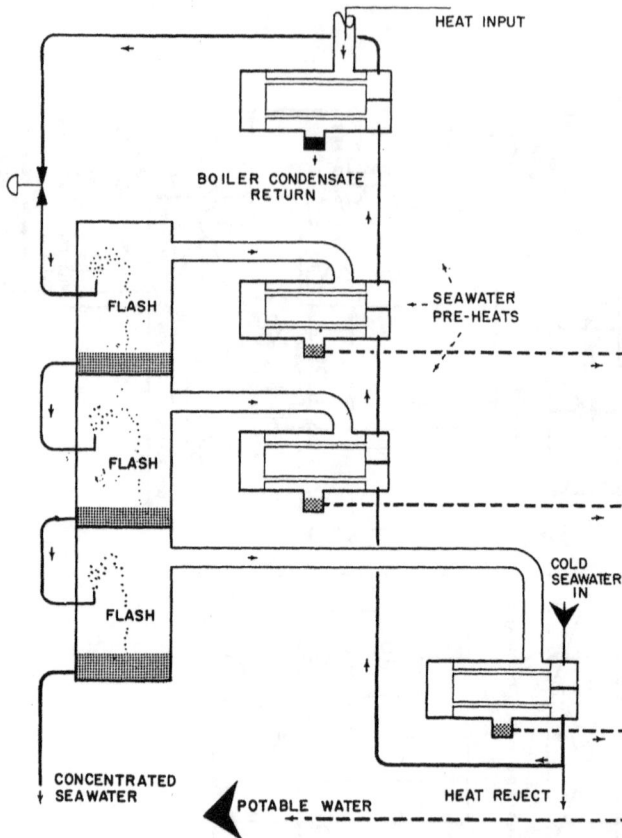

Courtesy of the Chicago Bridge and Iron Co.

FIG. 33.   BASIC MULTIPLE-FLASH DISTILLATION PROCESS

by piping the supply through the flash chambers starting at the low temperature end. The flashed vapor gives up its latent heat to the sea water as it condenses.

Vapor-Compression Distillation

In vapor-compression distillation, (Fig. 34) the steam produced in an evaporator is compressed and the heat thus released is used to raise the temperature of the raw liquid in the evaporator tubes. The compressed vapor, at its higher temperature, condenses when returned to the evaporator and gives up its latent heat to boil more salt water. The energy required for vaporization is that supplied to the vapor compression stage.

Thin films generated by wiping or rotating a surface covered by water

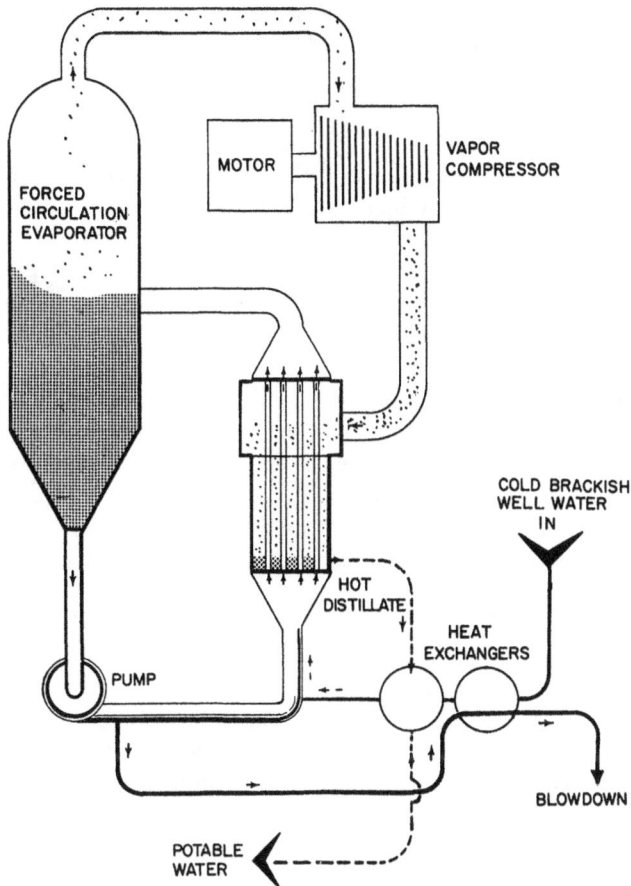

Courtesy of the Chicago Bridge and Iron Co.

FIG. 34.   BASIC VAPOR COMPRESSION DISTILLATION PROCESS

can be used to improve the efficiency of distillation processes. High conductivity results from maintaining both the evaporating sea water and the condensing vapor in the form of thin films. In one form, film thickness is governed by rotating a wiper on the evaporating side and by an extended fluted surface on the condensing side. General Electric has designed a 37,-000 gal./day thin-film pilot plant.

Solar Stills

Use of solar energy to distill water is an attractive prospect at first consideration. However, the area of absorbing surface required to collect the energy for vaporizing a gallon of water per day turns out to be large enough to make the construction and maintenance costs excessive except under certain restricted circumstances. In some geographical areas, solar stills having capacities of up to 100,000 gal./day might be practical. A type which has been used consists essentially of a shallow, glass-covered evaporating tray or basin. The bottom of the basin is black to absorb a maximum amount of heat which then distills some of the water into the space above the liquid. The vapor condenses on the sloping glass covers of the still and runs down to channels along the edges of the glass. Some development work has been done on improving this basic design. An experimental solar still at the University of Arizona separates the collection of energy, evaporation of salt water, and condensation of the product. Conventional tube and shell exchangers are used in the current model, but a modification constructed of thin plastic sheets is under development.

A type of evaporator not included in Dodge's classification is the fluidized bed still designed at Battelle Institute. Sea water, heated by condensing product vapor, is sprayed on a fluidized bed of salt particles. Evaporation occurs from the surface of the bed particles. The resulting steam is compressed, and dry salt is continually removed. Since concentrated brine is not produced, it is possible that scale would not be a problem in this apparatus.

Another approach to the scale problem has been developed by the W. R. Grace Corp. The principle involves the chemical removal of the scale-forming salts as a precipitate and the sale of the latter as a high analysis fertilizer. Wet-process phosphoric acid is added to the raw sea water to form insoluble magnesium and calcium phosphates (mostly magnesium ammonium phosphate and calcium hydrogen phosphate). Further processing of the precipitate would yield the fertilizer, sale of which would bring the net cost of the otherwise prohibitively priced technique to within an acceptable range.

FREEZING CONCENTRATION

Instead of separating pure water from brine by vaporization, it may be frozen and the ice crystals removed by filtration, centrifugation, etc (Fig. 35). In theory, freezing processes should be easier to engineer than distillation methods because scale is less of a problem and the heat transfer steps do not involve high temperatures. The principal difficulty which has been encountered seems to be the removal of entrained and adhering brine from the ice crystals. Usually, a portion of the purified water is used to wash the crystals after they have been freed of most of the brine by centrifugation or filtration. Latent heat of fusion is removed by flash

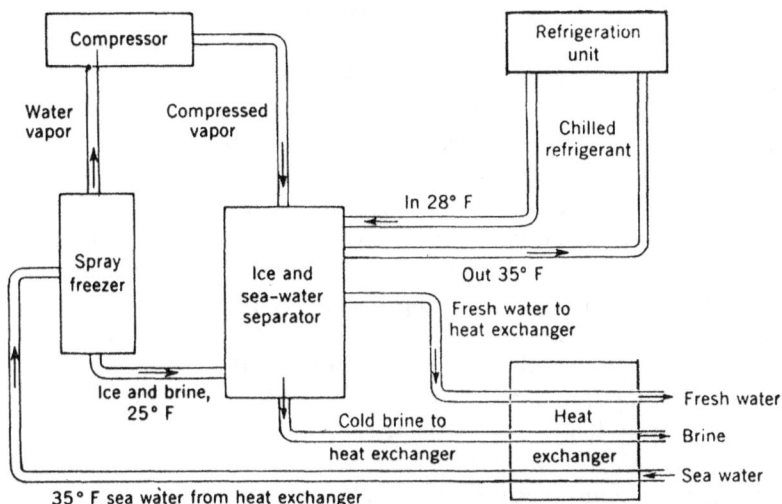

Courtesy of the Chicago Bridge and Iron Co.

FIG. 35. DIRECT FREEZING PROCESS IN WHICH DEAERATED SEA WATER IS PRE-
COOLED AND THEN FROZEN BY CHILLED REFRIGERANT

evaporation of precooled sea water at reduced pressures (3 to 4 mm. Hg)
or by vaporizing a refrigerant such as butane in direct contact with the
raw liquid.

SOLVENT EXTRACTION PROCESSES

The solvent extraction process seems to have some potential value for
desalting brackish waters containing 5,000 to 10,000 p.p.m. of salts. It involves transfering water from a solution where its activity is less than that
of pure water to another of still lower activity from which the activity
must then be raised to that of pure water. All solvents used so far extract
water from brine. An organic solvent which can dissolve water without
losing its identity as a separate phase is mixed with brine. After separating the solvent from the concentrated salt solution, water is released from
it by increasing the temperature slightly. Solvents most suited for use are
relatively low-molecular-weight organic compounds with one or more electronegative atoms such as nitrogen or oxygen. Highly substituted 5- and 6-
carbon secondary and tertiary amines or low-molecular-weight glycols and
glycol ethers have been recommended. The water obtained by solvent extraction frequently contains traces of the organic compounds which must
be removed by a separate heating or air-stripping stage.

ELECTRODIALYSIS

Desalination methods depending upon the properties of semipermeable membranes have achieved practical significance in recent years. Generally, the ion-selective film removes salt ions from the water, but some devices which have been suggested or investigated in a preliminary manner rely on membranes preferentially permeable to water. In the former case, the terms ion-selective or permselective are used to describe membranes which, if interposed between two electrolytes, show selective permeability toward ions of a particular sign. Such membranes are ion-exchangers in sheet form, a cation-exchange being selectively permeable to cations and an anion-exchanger showing the same behavior to anions. The two types are sometimes designated as "negative" and "positive," respectively, in accordance with the sign of the "fixed ions" in each case (Le Roux Malherbe and Mandersloot 1960). Electrodialysis is the phenomenon of ion transport through a membrane in which there is a driving force resulting from the application of an electrical field applied across the membrane. This is by far the most technically advanced process in the general category of membrane desalination.

There is an electrodialysis plant at Webster, South Dakota which takes in 48°F. brackish feed water containing 1,450 to 1,800 p.p.m. total dissolved solids and discharges about 250,000 g.p.d. of potable water having 300 to 350 p.p.m. salinity and a waste effluent of about 4,000 p.p.m. total dissolved solids. The plant consists of four stages, each stage composed of a stack of 216 membrane pairs. Membranes are held 0.030 in. apart by woven polyvinylchloride separators in polyester-lined steel frames. The effective area for ion transfer is about 10.4 sq. ft. per membrane. Cost of producing the potable water is less than $1 per thousand gallons. Electrodialysis is not considered to be competitive for processing waters higher in total dissolved solids than 10,000 p.p.m.

Figure 36 illustrates the principal of desalting by electrodialysis. In elementary form, a cell consists of an anode, a cathode, and a series of compartments separated by membranes which are permeable either to sodium ions or to chloride ions. The two types of membranes alternate in regular sequence across the cell. Brine is fed into the cell and the salt ions tend to migrate to the appropriate electrode. However, they are restricted in their travel by the membranes. As a result, the brine becomes more concentrated in some of the compartments while the salt concentration decreases in the others.

Ion-exchange techniques are sometimes listed in the category of membrane processes, but they are, in reality, more nearly analogous to chemical methods in which the ions are precipitated by forming insoluble compounds. A bed of ion-exchange resin takes up salt from the saline water

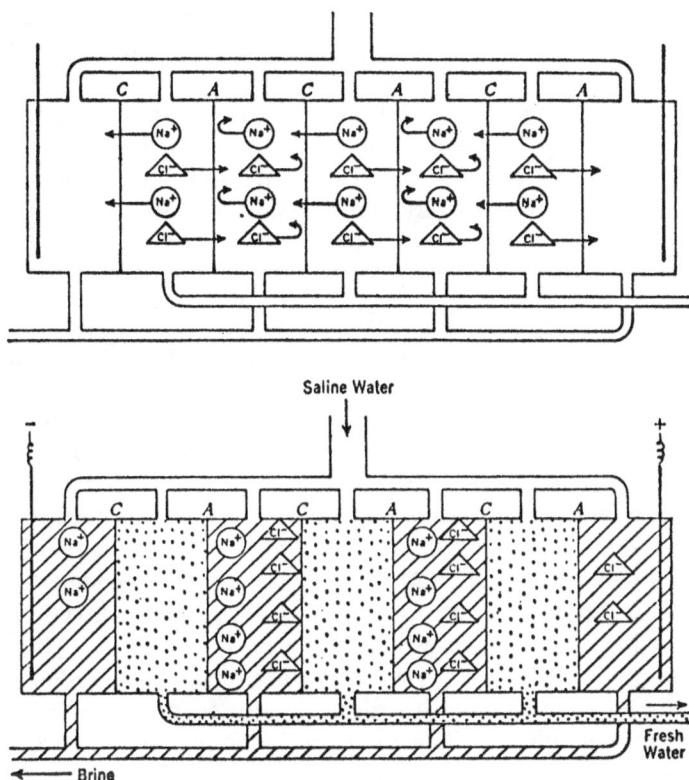

Saline Water

Fresh
Water

Brine

From Sieveka (1960)

FIG. 36. PRINCIPLES OF THE ELECTRODIALYSIS PROCESS

The top unit, with the current turned off, is filled with saline water.
When the current is turned on, anions and cations become concentrated in alternate compartments, and the water in adjoining
compartments becomes depleted of salt ions. The vertical lines
labeled *A* or *C* represent membranes: *A*, the anion-permeable
membrane; *C*, the cation-permeable membrane.

feed and the effluent is used directly. The resin is then regenerated by a
strong solution. Procedures based upon this principle are not considered
economically feasible because of the expense of the regeneration solutions.
Work is underway at Ionac Chemical Company on means for regenerating
resins with carbon dioxide under pressure of about ten atmospheres. Presumably, costs could be reduced to an attractive level through this approach.

Waste disposal problems created by conversion of saline waters are of unusual types and are not always given adequate emphasis by workers who publish in this field. These problems are connected with the ultimate disposal (not treatment) of exceptionally large quantities of waste concentrate. Some methods, such as injection into porous strata of rock, are relatively cheap, but, where these cannot be applied, alternate methods may cost many times as much as conversion itself (Koenig 1959). Concentrate disposal is less difficult for plants located near a seashore, of course.

REVERSE OSMOSIS

The process of reverse osmosis has received a great deal of favorable publicity in recent months. In the first few years of intensive research into desalination techniques it was given little consideration because distillation or ion-exchange techniques were based on principles that were better understood from an engineering standpoint and seemed to offer more chances of prompt practical success. Reverse osmosis is astonishingly simple, in principle. In the system pure-water: osmotic-membrane: saline-water, the pure water will tend to pass into the saline compartment (of course, water molecules are actually passing in both directions at all times). The tendency is measurable as a pressure increase in the saline water compartment. By applying a pressure to the salt solution which is greater than the osmotic pressure, the migration tendency will be reversed and a net gain of volume will be observed in the pure water compartment. The osmotic pressure when sea water is involved is about 350 lb./sq. in.

According to Sieveka (1960), the best practical membrane of several investigated by a University of Florida group was cellulose acetate, which reduced the salt content of water forced through it by about 95%. Improvements have since brought the reduction to 99%. At 600 lb./sq. in. the flow rate through this particular film of six microns thickness was 0.11 ml./sq. in./hr., i.e., 5.84 gal./day through each square yard of membrane. The film failed after 74 ml./sq. cm. were treated and it was concluded that both flow rate and life of the film were unsatisfactory for a practical water-producing process.

An apparent advantage of reverse osmosis is elimination of the need for expensive heat-generating and transfer equipment, since no phase change is involved. Energy requirements would be relatively low if a substantial part of the power required for applying pressure were recovered by a turbine after the water passes through the membrane. Sieveka calculated that under conditions of 600 p.s.i. pressure and with 25% of the water passing through the membrane before the remainder is rejected, 12.8 kw.h. of energy will be required per 1,000 gal. of fresh water produced.

The main difficulties revolve around the development of satisfactory equipment. A successful membrane must have considerable mechanical strength, although it can be supported by porous porcelain plates or other means. It must also be highly selective, passing water freely while retaining all of the several ionic species in the solution.

Aerojet-General Corp. is constructing a 1,000 gal./day reverse osmosis pilot plant at Newport Beach, Calif. Pressures of about 1,500 p.s.i. will be required. The cellulose acetate membranes will be about one mil. thick and are to be modified by treatment with magnesium perchlorate to increase the rate of water flux. Preliminary tests with an input of water containing 500 p.p.m. dissolved salt yielded fresh water with 200 p.p.m. solute at a flow rate of 60 gal./hr. The type of cell used in the Aerojet-General process is diagrammed in Fig. 37. The cell is divided into three modules,

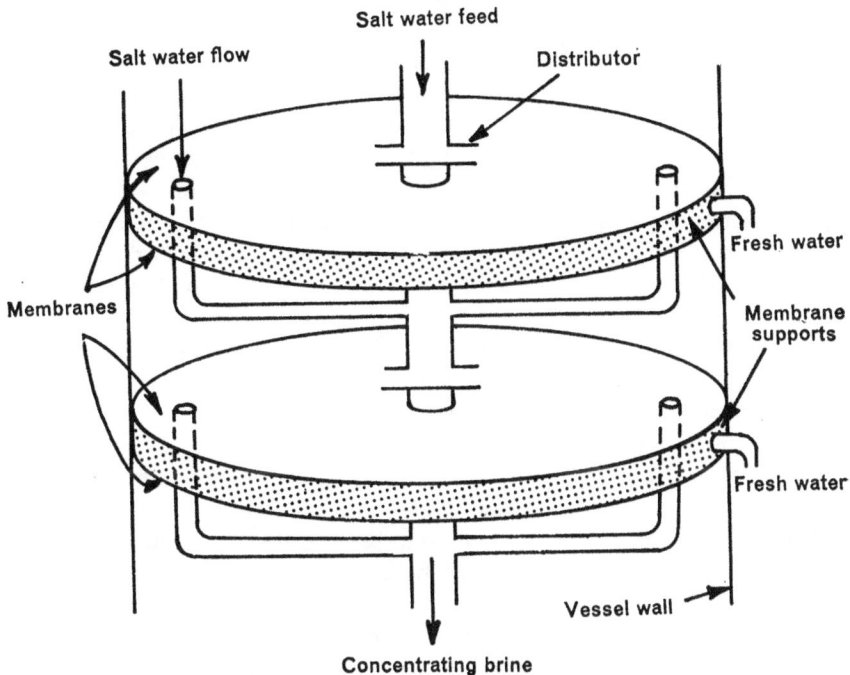

Courtesy of Aerojet General

FIG. 37. DIAGRAM OF CELL ARRANGEMENT FOR DESALINATION BY REVERSE OSMOSIS

each with 15 desalination plates. The plates are of laminar construction, with a membrane on top, a 0.5 in. thick support plate of glass fiber-reinforced epoxy laminate in the center, and a second membrane on the bottom. Filtered sea water at a pressure of 1,500 p.s.i.g. is supplied to the cell by a triplex piston-type pump. The purified water is collected from the support area (center of the cell sandwich) and a concentrated (about 45%) brine is drawn off at atmospheric pressure.

USE WITHOUT DESALINIZATION

Although there are problems involved in doing so, sea water can be used for certain culinary purposes without desalting it. The Armed Forces have conducted tests to determine the suitability of sea water as an ingredient in bread doughs. It was found that fairly respectable experimental loaves differing slightly or not at all from control bread in flavor could be made if the ingredient salt was omitted when sea water was used in place of fresh water. Similar arguments could be applied to the use of sea water in other bakery foods.

If sea water is used at an absorption of 60%, then the sodium chloride contributed by it to the dough is about two per cent (flour weight basis) which is actually a bit lower than the amount of ingredient salt usually added to bread doughs. It is certainly conceivable that purified but not desalinated sea water could be used for cooking vegetables and the like (rice, potatoes). Although the concentration of salt is somewhat greater than that normally added to the cooking water, most of this would be discarded after cooking is completed. Significant changes in flavor and perhaps in texture might occur due to the ionic constituents other than sodium and chloride, and scaling of cooking vessels might create problems.

It is not likely that fruits could be cooked in sea water or that beverages such as coffee or tea could be made from it without the introduction of some unpleasant and atypical flavors.

Meats and combination dishes such as stews also offer good possibilities for the utilization of sea water as a cooking medium or ingredient. Of course, the supply of two kinds of water to homes would create very difficult distribution problems, and the only location likely to prove practical for use of sea water in cooking is on board ship or in dire emergencies.

BIBLIOGRAPHY

ACKERMAN, E. A. 1963. Reasons for research and development on water desalting. Natl. Acad. Sci. Natl. Research Council Publ. 942, 5–16.
ANON. 1956. Fresh water from saline sources. Bull. 3. Ionics, Inc., Cambridge, Mass.
ANON. 1960. Saline Water Conversion. Advances in Chem. Ser. No. 27.
ANON. 1961. What price water—salt or fresh? Am. Water Works Assoc. News Bull. July 14, 1961.
ANON. 1962. Fresh Water from the Sea. Dechema Monograph. 26.
ANON. 1964. News release. Chem. Eng. News 42, No. 11, 21.

BARDUHN, A. G. 1963. Present and future research on the hydrate process. Natl. Acad. Sci. Natl. Research Council Publ. *942*, 121–133.

BARDUHN, A. G., TOWLSON, H. E., and YU, Y. C. 1962. The properties of some new gas hydrates. A. I. Ch. E. Journal *8*, 176–183.

BRADT, D. M. 1960. Distillation processes. J. Am. Water Works Assoc. *52*, 574–584.

BRICE, D. B., and TOWNSEND, C. R. 1960. Sea water conversion by the multi-stage flash evaporation method. Adv. in Chem. Ser. No. 27, 147–155.

CARTIER, R., PENDZOLA, D., and BRUINS, P. F. 1959. Particle integration rate in crystal growth. Ind. Eng. Chem. *51*, 1409–1414.

CARY, E. S., ONGERTH, H. J., and PHELPS R. O. 1960. Domestic water supply demineralization at Coalinga. J. Am. Water Works Assoc. *52*, 585–593.

CHIRICO, A. N. 1963. Fresh water from saline waters for food processing. A I. Ch. E. Food Proc. Symp. *1963*. Preprint 22.

CLAUSSEN, W. E. 1949. Suggested structures of water in inert gas hydrates. J. Chem. Phys. *19*, 259–270.

COHAN, H. J. 1961. Electrodialysis—equipment and membranes. Chem. Eng Progr. *57*, No. 2, 72–75.

COWAN, D. A., and BROWN, J. H. 1959. Effect of turbulance on limiting current in electrodialysis cells. Ind. Eng. Chem. *51*, 1445–1448.

CYWIN, A. 1958. Summary of industrial accomplishment and research in saline water distillation. Natl. Acad. Sci.–Natl. Research Council Publ. *568*, 29–34.

DODGE, B. F. 1962. Need for research in process involving a phase change. Natl. Acad. Sci.–Natl. Research Council Publ. *942*, 192–203.

DUMESNIL, G. 1958. Saline water conversion research in the world. Proc. Symp. on Saline Water Conversion *1957*, 3–11.

GEYER, J. C. 1963. Water conservation. Natl. Acad Sci.–Natl. Research Council Publ. *942*, 17–49.

GILLIAM, W. S. 1963. Research activities—Office of Saline Water. Natl. Acad. Sci.–Natl. Research Council Publ. *942*, 77–85.

HIMES, R. C., MILLER, S. E., MINK, W. H., and GOERING, H. L. 1959. Zone freezing in demineralizing saline waters. Ind. Eng. Chem. *51*, 1345–1348.

KARNOFSKY, G. 1961. Saline water conversion by freezing with hydrocarbons. Chem. Eng. Progr. *57*, No. 2, 42–46.

KOENIG, L. 1959. Economic boundaries of saline water conversion. J. Am. Water Works Assoc. *51*, 845–862.

LAUTH, H. 1961. Recent experiences with ion exchangers in the desalting of water. Wasser Luft Betrieb. *5*, 275–278.

LE ROUX MALHERBE, P., and MANDERSLOOT, W. G. B. 1960. The physical chemistry of ion-selective membranes. *In* Demineralization by Electrodialysis. Edited by J. R. Wilson. Butterworths, London.

MCRAE, W. A. 1962. Liquid-phase permeation through membranes. Natl. Acad. Sci.–Natl. Research Council Publ. *942*, 255–269.

MEYER, K. H., and STRAUS, W. 1940. The permeability of membranes. Passage of electric current across selective membranes. Helv. Chim. Acta *23*, 795–800.

MORGAN, W. H. 1961. Augmentation of supply by sea-water distillation. J. Institution Water Engineers *15*, 266–298.

MURPHY, G. 1957. The Osmionic Process. Office of Saline Water Progr. Rept. No. *14*.

RAPIER, P. M., WEINER, S. A., and BAKER, W. K. 1963. Experimental studies of some basic parameters in electrodialysis. Calif. Univ. Water Resources Center Contrib. 70.

ROTHBAUM, H. P. 1958. Desalination energy factors. Chem. Eng. Data Series 3, 50–52.

SIEVEKA, E. H. 1960. Review of desalinization processes: membrane processes. J. Am. Water Works Assoc. 52, 556–568.

SKERRITT, D. E. 1963. A study of brackish water demineralization by ion exchange. Am. Chem. Soc. Div. Water Waste Chem. Preprints 1963, 215–222.

SPIEGLER, K. S. 1962. Salt-Water Purification. John Wiley and Sons, New York.

STANDIFORD, F. C., and BJORK, H. F. 1960. Evaporation of sea water in long-tube vertical evaporators. Advances in Chem. Ser. No. 27, 115–127.

STROBEL, J. J. 1963. Saline-water-conversion processes. Natl. Acad. Sci. Natl. Research Council Publ. 942, 86–106.

SVERDRUP, H. U., JOHNSON, M. W., and FLEMING, R. H. 1942. The Oceans. Prentice-Hall, Englewood Cliffs, N. J.

TRIBUS, M., and EVANS, R. B. 1963. The thermo-economics of saline water conversion. Am. Chem. Soc. Div. Water Waste Chem. Preprints 1963, 150–176.

V. MENTS, M. 1960. Water desalinization by electrodialysis. Ind. Eng. Chem. 52, 149–152.

WASSERMAN, W. J. 1963. Demineralization of saline water through pressurization cycles with ion-exchange materials. Am. Chem. Soc. Div. Water Waste Chem. Preprints 1963, 177–195.

WEBB, R. M., UNTERBERG, W., GREGSON, W., JR., and PERRINE, R. L. 1963. Saline water evaporation from wiped films. Am. Chem. Soc. Div. Water Waste Chem. Preprints 1963, 234–239.

WILSON, J. R. 1960. Demineralization by Electrodialysis. Butterworths, London.

WILSON, J. R., COOKE, B. A., MANDERSLOOT, W. G. B., and WIECHERS, S. G. 1960. The electrodialysis process. In Demineralization by Electrodialysis. Edited by J. R. Wilson. Butterworths, London.

CHAPTER 9

Determining the Moisture Content of Foods
and Ingredients

INTRODUCTION

The determination of moisture content was probably one of the first analytical procedures applied to foods and it is today undoubtedly the most common. Such techniques have prime economic significance because they indicate the proportion of a "diluent" present in a foodstuff and they have fundamental importance in that the amount of a substance which often affects the sensory properties and storage stability of a food is measured.

Much effort has been devoted to the development of techniques for moisture analysis which: (1) reduce the amount of work the analyst must expend per determination; (2) reduce the skill required for performing the analysis; (3) reduce the sources of error; (4) increase the precision; and (5) increase the speed with which results become available. Other considerations sometimes motivating such development activities include: (1) reduction of sample size and (2) decreasing the equipment cost.

THERMAL METHODS

Thermogravimetric methods predominate in this field because the principle involved is relatively straightforward, the time required for each analysis is usually not excessive, and only moderate skill is required of the operator. The latest edition of the Official Methods of the Association of Official Agricultural Chemists (Anon. 1960) lists thermogravimetric tests for moisture or total solids content of the following foods, beverages, and feeds: (1) baked products other than bread; (2) bread; (3) brewers' grains; (4) butter; (5) cacao products; (6) canned vegetable products; (7) cereal adjuncts; (8) cheese; (9) coffee, roasted; (10) condensed milk, sweetened; (11) confectionery; (12) cordials and liqueurs; (13) cream; (14) dressing for foods; (15) dried fruits; (16) dried and malted milk; (17) eggs and egg products; (18) evaporated milk; (19) fig bars and raisin-filled crackers; (20) fish; (21) flour; (22) fruits and fruit products; (23) gelatin dessert powders; (24) ginger extract; (25) grain and stock feeds; (26) grain products; (27) highly-acid milk by-products; (28) honey; (29) hops; (30) ice cream; (31) frozen desserts; (32) lemon, orange, and lime extracts; (33) macaroni, egg noodles, and similar products; (34) malt; (35) maple products; (36) marine products other than fish;

(37) mayonnaise and salad dressings; (38) meat and meat products; (39) meat extracts; (40) milk; (41) mineral waters; (42) non-alcoholic beverages and concentrates; (43) nuts and nut products; (44) oils, fats, and waxes; (45) oysters, raw; (46) plants; (47) prepared mustard; (48) salt; (49) soybean flour; (50) spices; (51) starch conversion products such as corn syrups and refined and crude corn sugars; (52) starch dessert powders; (53) sugars and sugar products; (54) tea; (55) vinegar; (56) wheat, rye, oats, corn, buckwheat, rice, and barley and their products; and (57) yeast.

Only a handful of official methods rely on non-thermogravimetric procedures. Among these are thermovolumetric (toluene distillation) methods for grains and indirect techniques such as refractometry or density measurements for sugar syrups, jellies, etc.

In spite of the popularity of the "weigh, dry, and weigh" methods, it has long been recognized that they have serious disadvantages from the fundamental standpoint. Volatile substances other than water can be driven off, even under rather mild conditions (many workers have shown this—see, for example Schäfer and Seibel 1957), chemical reactions leading to the creation of water or other volatile substances can occur, and fractions of the moisture present are bound with varying degrees of tenacity by the hydrophilic substances present.

Practical difficulties of maintaining control of the drying conditions also beset the analyst using oven methods of moisture determination. Köster (1934) and Oxley and Pixton (1961) have listed some of the conditions whose variations affect moisture results as: (1) drying time; (2) average temperature of the drying atmosphere; (3) temperature of the exit air; (4) air pressure in the drying space; (5) relative humidity of the drying atmosphere; (6) relative humidity of the exit air; (7) air speed through the oven; (8) depth of the samples; (8) particle size of the samples; (9) number of samples; (10) type of oven; (11) position of the samples relative to each other; (12) barometric pressure; (13) uncontrollable variations in the temperature; (14) variations in the weights of the drying dishes; and (15) variations in sample preparation techniques. Not all of these will affect the accuracy in every method, but most will be found to influence results in the usual procedures. But in spite of the formidable list of disadvantages, drying techniques will continue to be popular because they can be performed with simple, inexpensive equipment and can be made to yield results of sufficient reproducibility for most purposes.

Many specialized devices for thermogravimetric determinations have been developed with the intention of decreasing the demand on the analyst's time or of increasing the rapidity with which results become available. These are too numerous to review in detail, but include more common devices such as the Brabender moisture oven which uses dishes of uniform tare in connection with a forced-air oven containing a built-in balance (Fig 38). Equipment using infrared heaters to speed up drying has been developed (Lincoln *et al. 1954*). A thorough review of these methods was given by Mitchell and Smith (1948).

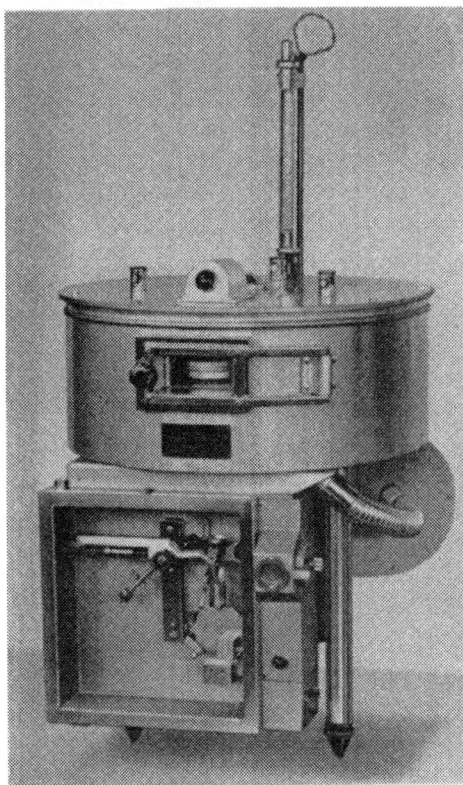

Courtesy of C. W. Brabender Instruments, Inc.

Fig. 38.   Brabender Moisture Oven

A few thermovolumetric methods have been used for foods. These usually involve distillation of the water from the sample in contact with toluene or with a high-boiling hydrocarbon. The vapor is condensed and collected in a graduated tube. Toluene is particularly suitable because it forms an azeotrope boiling at 185°F., so that pyrolytic reactions leading to the creation of water are minimized. The azeotrope contains 79.8% toluene and 20.2% water. The upper layer in the collection tube will contain 99.95% toluene while the lower layer will contain 99.94% water, permitting the direct reading of water volume without the necessity for correcting for dissolved toluene, in ordinary work.

Liquids other than toluene have been used as heat transfer media and temperature control agents in distillation methods for moisture determination, as in the Brown-Duvel method for grain.

GAS CHROMATOGRAPHY

Gas chromatography is a general method suitable for separating from a mixture and identifying almost any substance which can be vaporized without decomposition. A gas chromatograph will consist of a tube, usually of several feet in length, filled with a granular support medium coated with a liquid partitioning agent. A liquid or gas mixture is introduced at one end of the tube, often by injection through a rubber or plastic diaphragm. If the mixture is liquid, it is vaporized by an external heat source. Vapors are moved through the column by an inert carrier gas such as helium.

The components of the mixture will migrate through the column at different rates depending mostly on their individual affinities for the partitioning agent and will emerge in fractions consisting of more or less perfectly separated compounds. Emergence of a fraction can be detected by methods which measure differences in heat conductivity, ionization potential, or other characteristics of the gases passing through the detector.

One common form of detector for gas chromatography is based on the principle of thermal conductivity. As the gas emerges from the column of adsorbent it passes over a fine, electrically heated wire. When a constant flow of carrier gas is passing over the wire, the rate of heat loss by the wire is constant. However, a change in composition of the gas stream causes a change in the loss of heat from the wire, thus changing the temperature and consequently the resistance of the wire. Changes so produced are amplified and plotted as a series of peaks on a chart. The area under the peak is proportional to the quantity of a component. Under a defined set of operational conditions, the retention time, i.e., the time the constituent requires to pass through the column, is characteristic of each compound. However, the retention times of different compounds may be the same within the limits of error of the measurement. Retention times are affected by temperature, rate of flow of carrier gas, partitioning medium, characteristics of support granules, length and cross-sectional area of the column, etc. Electronic integration of the curves is used in conjunction with standardization procedures to give a quantitative measure of the compound under investigation.

The water content of a gaseous or liquid mixture can be determined by gas chromatography, although difficulty in the form of excessive tailing of the water due to adsorption on the stationary or supporting phase is frequently observed. Bennett (1964) found that a commercially available

Teflon column was satisfactory for the quantitative determination of water in a variety of aqueous-organic systems. He used an F & M Model 500 gas chromatograph with thermal conductivity detector. The commercial column was four feet by one-fourth inch stainless steel tubing packed with Teflon powder impregnated with five per cent Carbowax 20-M. Other conditions were: column temperature 122°F., carrier gas 70 ml. of helium per minute, sample size six microliters. The maximum average deviation for duplicate determinations was ±0.10%.

Earlier, Carlstrom *et al.* (1960) had used a column packed with 30% polyethylene glycol of about 200 molecular weight on 20- to 30-mesh insulating type firebrick to determine small amounts of water in butane. A preliminary separation of the water was performed by passing the vaporized mixture through a trap containing the same type of packing as the column. Helium was passed through the column at a rate of about 100 ml./min. Under these conditions, retention time for water was 11.5 min. Samples were analyzed with a Perkin-Elmer Vapor Fractometer Model 154B using a two foot column maintained at 176°F.

Weise *et al.* (1964) developed a method for estimating the water content of materials such as food. First, the water is extracted with methanol. The methanol solution is then chromatographed on a commercial 30–60 mesh polytetrafluoroethylene coated with a polyethylene glycol partitioning agent. The method is highly specific and has the additional advantage of being largely instrumental. Weise *et al.* stated that there is good reason to believe that the water-extractable-by-methanol consists of all but that which is chemically combined.

Penther and Notter (1964) separated traces of water from hydrocarbon streams in a 0.25 in. diam. copper tube lined with 0.185 in. outside diameter by 0.125 in. inside diameter Teflon tubing. The packing was Fluoropak-80 coated with ten grams per 100 gm. of polyethylene glycol. Detection was with the Keidel cell (1959)—described in Chapter 10—which is specific for water under the conditions of use.

The procedure of Schwecke and Nelson (1964) employs a 0.25 in. outside diameter by 5 ft. long column containing Fluoropak-80 coated with ten per cent of Carbowax 400. Water has a low retention time in this column, allowing completion of the separation in about five minutes. Secondary butanol is used as an internal standard and methanol is used to extract water from the product. The standard deviation appeared to be no more than about ±0.10% for substances with 9 to 22 % moisture. The column temperature was 248°F. (injection port at 302°F.) and the carrier gas, helium, flowed at 65 ml./min. Sample size was two microliters. A hot-wire detector was used.

Stein and Ambrose (1963) separated water on a column packed with Fluoropak-80 which had been coated with LAC2R446 (see their article for further identification) and 0.4% phosphoric acid. The water was quantitatively determined by measuring its absorption in the near-infrared. Results so obtained were considerably lower than those obtained with the Karl Fischer titration, probably due to the reaction of the product (aluminum aspirin) with the titrant in the latter procedure.

It is quite evident that gas chromatography has a definite place as a tool for moisture determination. Its advantages over other methods of analysis are its extreme sensitivity and its excellent specificity under ideal conditions of analysis. Disadvantages are the special preparation of sample usually required and the initial cost.

NUCLEAR MAGNETIC RESONANCE

Bloch *et al.* (1946) and Purcell *et al.* (1946) discovered in 1945 that hydrogen nuclei can absorb radio-frequency energy. One of the uses to which others adapted this discovery was the determination of the moisture content of substances (Shaw and Elsken 1950).

A description of atoms which is consistent with the data accumulated by physicists pictures the nuclei of atoms as rotating about their central axes. Since these nuclei possess an electric charge and the rotation of a charged body creates a magnetic field, the nuclei behave like spinning permanent magnets. When an external magnetic field is applied, the nuclei tend to orient their axes in fixed directions congruent with the external field. For hydrogen nuclei, there are two possible orientations of the nuclear axes with respect to the external magnetic field. Each of these positions corresponds to a different and characteristic energy level. By varying the external field, the orientation of the axes can be switched back and forth between the two positions.

If the nuclei lie in the field of a resonant radio-frequency generator which provides waves of a frequency concordant with the strength of the magnetic field (42.6 mc./sec. at 10,000 gauss, for example), measurable quantities of radiant energy will be absorbed by the nuclei as they change to the higher energy level. The amount of energy absorbed is related to the number of nuclei present within the field. Thus the phenomenon permits a quantitative estimation of the number of hydrogen nuclei in a sample. In most foods, the amount of hydrogen nuclei present in water is greatly exceeded by the number of hydrogen atoms in the proteins, carbohydrates, fats, and other components of the material. Therefore, a straightforward determination of the quantity of hydrogen present in a food would not be of value as a basis for estimating the moisture percentage. However, instruments employing the nuclear magnetic resonance principle can be constructed so as to differentiate between tightly bound hydrogen atoms in large molecules and hydrogen atoms loosely bound in water. The signals which are obtained depend upon the force with which the water is bound to other compounds in the food so that highly polar materials such as gelatin give weak signals even at ten per cent moisture, while sugar will give a detectable signal when only a few hundredths of one per cent water are present. Wheat flour gives a useful impulse down to about 7.5% moisture.

In determining moisture content of a sample by the nuclear magnetic resonance method, a small amount of the substance in a glass tube or similar container is placed inside a radio frequency coil which is situated between the poles of a powerful permanent magnet. The frequency of the radio waves is held constant while the magnetic field is varied over a relatively narrow range in a predetermined pattern by applying a small current to windings on the permanent magnet (Fig. 39). The absorption of radio frequency is plotted against magnetic field strength. The amount of absorption is related to the number of resonant nuclei present in the sample.

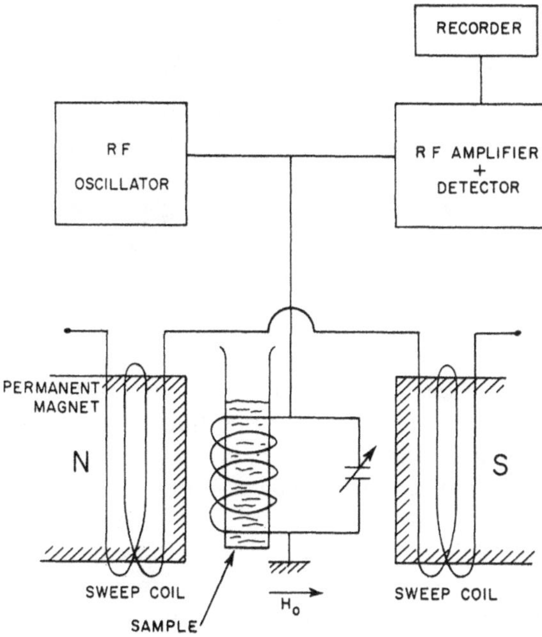

Courtesy of Varian Associates

FIG. 39.   BASIC NMR ANALYSIS CIRCUIT

The Varian PA-7 NMR Process Analyzer is an instrument operating on the foregoing principles. The two major parts of the equipment are a 1717 gauss permanent magnet and an electronic control console. A plug-in chassis containing the radio frequency amplifier and detector is located inside the magnet yoke. Provision is made for inserting the sample container within the radio frequency coil (Fig. 40). The recorder and con-

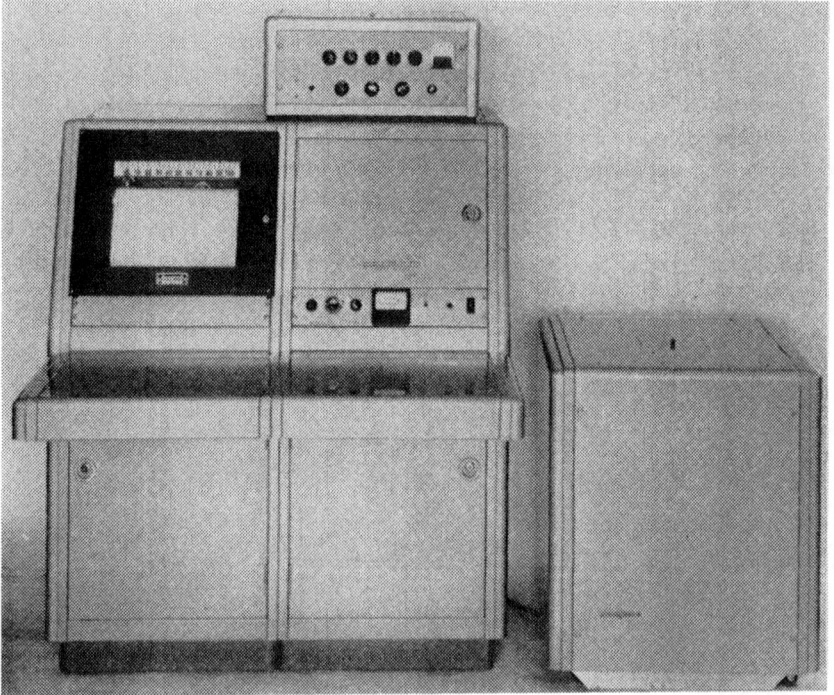

Courtesy of Varian Associates

FIG. 40.   THE VARIAN PA-7 NMR PROCESS ANALYZER

trols are located in the console cabinet. Selector switches permit adjustment of the radio frequency level, sweep time, sweep width, modulation amplitude, signal gain, and filter-time constant. The magnetic field is varied linearly across the nuclear resonance values by a small current applied to the sweep coils by a sweep generator.

When a fixed radio frequency value is used in a device of this type, an absorption spectrum will be obtained. The Varian apparatus records a derivative absorption curve which has experimental advantages over the absorption spectrum. The curve measurement of importance to the operator making moisture determinations is the peak-to-peak height. A calibration curve obtained by analyzing samples of known moisture content must be drawn up for each type of material which will be investigated because of the differences in binding strength previously mentioned. These curves are frequently nearly linear over a considerable range of moisture levels but tend to depart from the straight-line relationship at low water percentages (see Fig. 41).

Among the advantages claimed for the nuclear magnetic resonance method of moisture determination are: (1) it is rapid in comparison with standard techniques, the actual measuring time being about 30 sec.; (2) it is non-destructive; (3) its precision is equivalent to that of standard techniques; and (4) sample preparation is minimal and most solids may be used "as is" (Rubin 1958).

Shaw and Elsken (1950) were perhaps the first to publish results of moisture determinations of foodstuffs obtained by nuclear magnetic resonance techniques. They reported experimental results for fresh and dehydrated apple and potato which implied that the precision of the test used by them was about one per cent. Shaw *et al.* (1953) indicated the principal limitation on the precision of their procedure was the lack of reproducibility of the packing and the non-uniformity of the test specimens. Shaw and Elsken (1956) found that the water content of potato tissue could be determined provided corrections were made for soluble solids, the hydrogen nuclei of which contributed to the nuclear magnetic resonance. Palmer and Elsken (1956) further explored the contribution of hydrogen nuclei of solutes to the observed signal obtained from sugar solutions, milk, and apple juice concentrates and showed that they absorbed in a manner indistinguishable from hydrogen nuclei of water in the high moisture region under the conditions of operation of the low resolution spectrometer described by Shaw and Elsken. Conway *et al.* (1957) obtained favorable results in applying nuclear magnetic resonance to the determination of moisture in candies.

A more advanced instrument was available to Miller and Kaslow (1963) who used it to determine moisture contents of wheat, flour, doughs, and dried fruits. All data gave essentially linear calibration curves when integrator read-out measurements (i.e., areas) were plotted against moisture as determined by standard air-oven methods. Area measurements yielded more nearly linear calibration curves than did peak-to-peak measurements, particularly at low moisture levels. Curvilinear relationships existed between nuclear magnetic resonance and the percentage of moisture, generally when the latter was less than six per cent. This effect is due to an interaction between the adsorbed water and the adsorbent. The curvilinear portion was usable, however, down to about four per cent moisture. Below four per cent, the signal remained essentially constant.

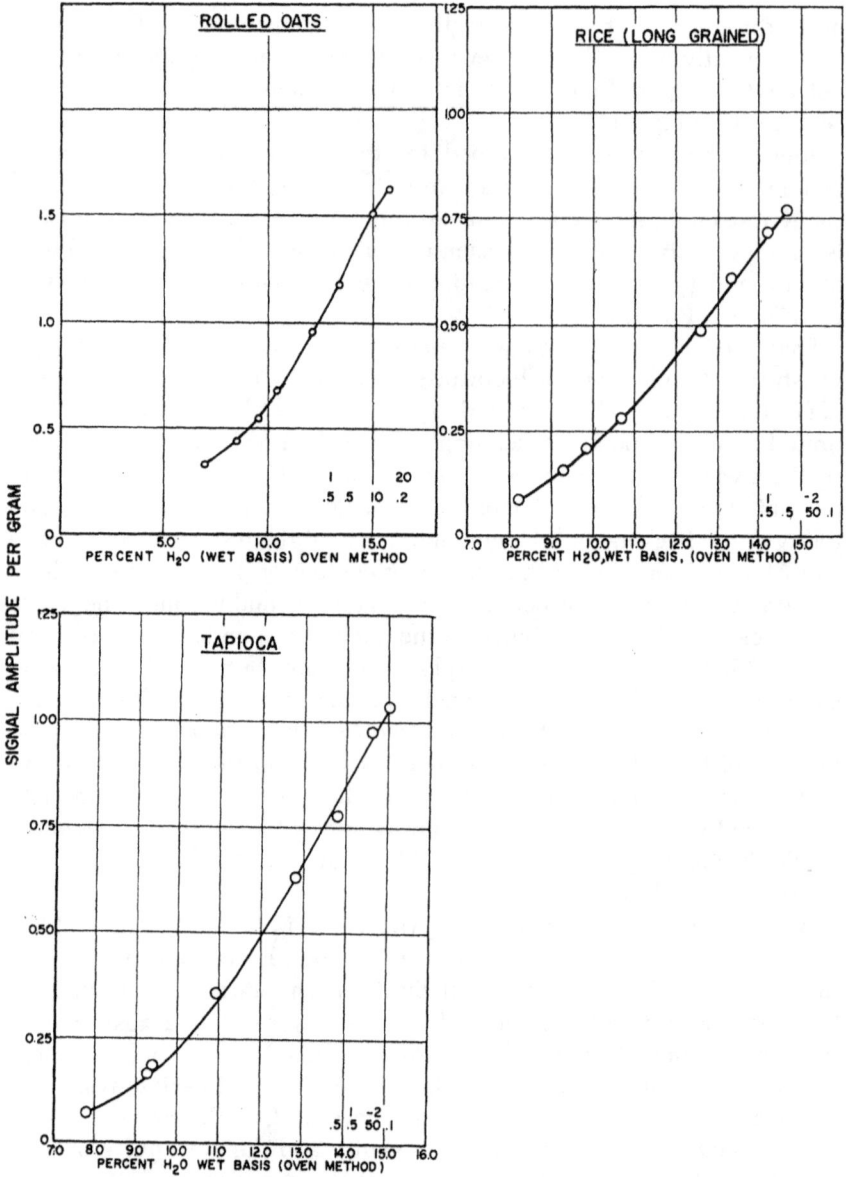

FIG. 41. CALIBRATION CURVES FOR THREE COMMODITIES

Courtesy of Varian Associat.s

CONDUCTIVITY

The resistance of food products to the passage of an electric current varies with the amount of water present in the substance. This effect has been used as the basis for the design of several "moisture meters." Such devices are extensively used in testing grains, where their speed of operation is a great asset. In the Tag-Heppenstall (Weston) meter, kernels of wheat or other grains complete an electric circuit as they are passed between rotating corrugated cylinders. The resistance is indicated on an ohmeter and the moisture content can be determined from this reading by consulting a calibration chart. In spite of the theoretical objections which might be raised, the equipment appears to work perfectly satisfactorily in practice.

Other moisture meters relying on the conductivity principle have means for compressing the sample in a chamber of definite size. It is obvious that the area of contact between the conducting kernels has an important bearing on the conductivity of the mass, and it is necessary to keep this factor as constant as possible. These meters are said to give readings accurate to at least 0.5% over the range of 12 to 20% moisture. Christensen and Linko (1963) evaluated a number of these meters.

A disadvantage of the conductance-type moisture meters is that their accuracy is strongly dependent upon the existence of a uniform distribution of moisture within the kernels of grain. If the kernels have recently been dried, incorrect low readings may be obtained because the surfaces of the particles are abnormally dry in respect to the water content of the grain as a whole. Similarly, if the wheat has been recently moistened (as in preparation for milling) unduly high readings will be observed because of a higher level of water in the surface layers than in the interior. Mixtures of wet and dry grain may also fail to give correct readings (Zeleny 1960).

An indirect conductivity method involving the use of alcohol-acetone-water-sodium chloride systems formed by extraction of the sample was described by Hancock and Burdick (1956). The basis of their method is that conductivity of the system will be proportional to the sodium chloride concentration which, in turn, will be proportional to the water concentration. Use of a large excess of sodium chloride tends to mask the effect of electrolytes in the material being tested. When cottonseed meal was tested by this method, a standard deviation of ±0.2% was indicated and the correlation coefficient of this method with an oven method was 0.997%.

CAPACITANCE

The dielectric properties of a foodstuff will depend mainly on its moisture content so that the capacity of a condenser in which the food is a dielectric will vary accordingly. Meters using this principle have been developed and are commercially available. In some, the circuits are designed to use the change in impedance resulting from the change in capacitance. The impedance decreases with increasing moisture content of the substance.

Capacitance-type meters appear to be less subject than conductivity devices to errors caused by uneven moisture distribution in the sample. They are also usable over a wider range of moisture content. However, they are probably no more accurate than conductance meters in testing the normal run of commercial grain, according to Zeleny (1960). In general, replicability is better with the latter type of instrument.

COULOMETRIC METHODS

Keidel (1959) described a specially designed cell which could be used to determine the water content of a sample by quantitative hydrolysis. He proposed the method especially for gases having water concentrations of 1 to about 100 p.p.m. but suggested procedures for adapting it for determination of moisture in solids and volatile liquids as well. The analysis is accomplished with a specially designed electrolysis cell in which all entering water is continuously and quantitatively absorbed and electrolyzed to hydrogen and oxygen. The electrolysis current is used as an indication of water content. In accordance with Faraday's law, the electrolysis of one gram mole of water, i.e., 18.02 gm. requires 193,000 coulombs. The mass rate of entry of water into the cell is the quantity actually measured, but concentration can be determined if the flow rate and pressure of the gas stream are known.

Keidel's cell contains two platinum wire electrodes 16 ft. long. The wires are wound in a double helix evenly spaced at 0.003 in. An interelectrode coating of essentially anhydrous phosphorus pentoxide (with a measured resistivity of 10^{10} ohm cm.) is obtained by filling the cell with phosphoric acid, allowing it to drain, applying voltage to the system, and electrolyzing to dryness. The electrode is encased in a glass tube through which the stream of gas flows. When water is absorbed from the gas by the phosphorus pentoxide matrix, current can flow between the platinum wires. The current hydrolyzes the water and converts the metaphosphoric acid created by the absorption process back to phosphorus pentoxide. This becomes a continuous process operating at an equilibrium level. The current flowing between the electrodes is proportional to the rate of water absorption in the cell.

As discussed in a preceding section of this chapter, Penther and Notter (1964) have combined gas chromatographic procedures and the Keidel cell (operating as the detector) in a device applicable to the determination of moisture in foods.

A commercial coulometric moisture analyzer using a similar electrolytic cell is offered by the Consolidated Electrodynamics subsidiary of Bell and Howell. The sample is heated to drive off the water which is then picked up by a continuously flowing stream of externally dried nitrogen. The carrier gas and the water vapor pass through the electrolytic cell. A schematic diagram of the instrument is shown in Fig. 42.

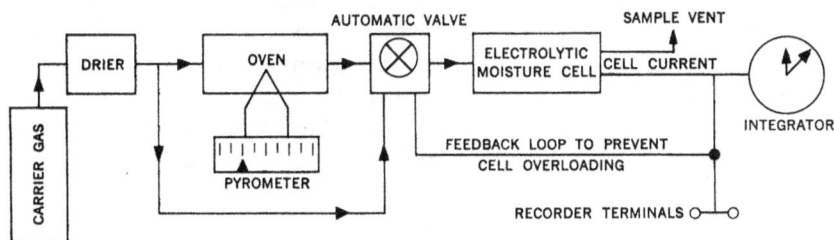

Courtesy Consolidated Electrodynamics, Division of Bell and Howell

FIG. 42. PRINCIPLES OF OPERATION OF THE CEC TYPE 26-321 SOLIDS MOISTURE ANALYZER

CHEMICAL METHODS

McNeil (1912) originated a method for determining moisture which relied upon the volumetric measurement of the acetylene gas produced by the interaction of calcium carbide and the substance under investigation. He stated that the procedure was applicable to such materials as flour, soap, leather, powder, vanilla beans, sawdust, butter, and fruit juices. Blish and Hites (1930) refined the calcium carbide method for flour moisture. They recommended mixing a one gram sample of flour with calcium carbide in a closed dry system connected with a mercury manometer. Pressure produced by the resultant acetylene is proportional to the water content of the sample. An appreciable fraction of the flour moisture determined by drying for one hour at 266°F. does not react with the carbide under test conditions. However, the quantity of non-reacting water appears to be reasonably constant among flours of different types, grades, and moisture contents. The unreacting moisture is about 4.55% of the total flour weight.

The carbide method never seemed to attain general currency, but in 1935 Karl Fischer (1935) published a method based on the chemical reactivity of water which was destined to be widely adopted. Essentially, Fischer's procedure involved a direct titration of the wet material with a solution of iodine, sulfur dioxide, and pyridine in methanol. The reagent serves as its own indicator. Later, Seaman *et al.* (1949) modified the method by using two solutions to retard deterioration of the reagents. They used a solution of pyridine and sulfur dioxide in methanol to dissolve or suspend the sample and a solution of iodine in methanol for titration.

According to Fischer, water reacts with the other components of the system according to the equation:

$$2H_2O + SO_2 + I_2 \rightleftharpoons H_2SO_4 + 2HI$$

The pyridine fixes the HI and H_2SO_4, and the amount of iodine which has been used up can be read directly by titrating with a standard solution of appropriate concentration. The color change due to the complete reaction of the dissolved iodine is the end point of the titration.

Smith *et al.* (1939) offered a somewhat different explanation of the course of the reaction. According to these workers, there are two distinct steps:

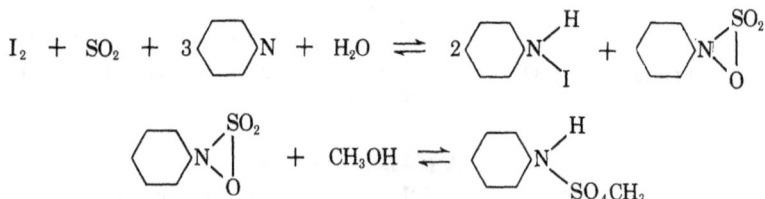

$$I_2 + SO_2 + 3\,C_5H_5N + H_2O \rightleftharpoons 2\,[C_5H_5N{\cdot}HI] + [C_5H_5N{\cdot}SO_3]$$

$$[C_5H_5N{\cdot}SO_3] + CH_3OH \rightleftharpoons [C_5H_5N{\cdot}H{\cdot}SO_4CH_3]$$

The first reaction is specific for water but the theoretical efficiency of one mole of water absorbed per mole of iodine used is rarely obtained because of side reactions. These authors recommend the Fischer reagent for determining water in alcohols, hydrocarbons (including saturated, ethylenic, and aromatic types), carboxylic acids except formic acid, and esters. Titration of aldehydes and ketones with the regular reagent is unsatisfactory due to the formation of acetals and ketals with the large excess of methanol and the resultant liberations of water. By greatly reducing the methanol content and greatly increasing the pyridine, a modified reagent is obtained which reacts normally in the presence of ketones provided large amounts of lower alcohols are not present.

The Fischer reagent can be standardized against water or a methanol solution of water, but water has the disadvantages of relatively high volatility and an unfavorable gravimetric factor. For routine analysis, it is desirable to use as the primary standard a stable crystalline hydrate of accurately known water content. These substances are more easily handled than water and have more favorable gravimetric factors. Some of the substances which have been recommended are ammonium oxalate monohydrate, sodium acetate trihydrate, and citric acid monohydrate. Neuss *et al.*

(1951) advocate the use of sodium tartrate dihydrate because the water content is stable at extreme humidities and it contains very close to the theoretical amount of water.

The Karl Fischer titration has been used for routine moisture determination of a wide variety of foods. Examples of published procedures are Seibel and Bolling (1958) for grain products, and Sloman *et al.* (1954) for chocolate.

INDIRECT METHODS

Although refractometry and methods which rely on the depth of submergence of a weighted bob (i.e., the density of a solution) are usually regarded as indicating the amount of soluble solids (generally considered as all sugar), they could equally well be thought of as means for determining the moisture content of the system. There are many theoretical objections which could be advanced against the use of these methods, but they are, nonetheless, of considerable practical value (see Fig. 43).

Refractometry is very commonly used to obtain an estimate of total soluble solids in jellies, jams, and syrups of various kinds. In these products, sugar and water are regarded as the only constituents *significantly* affecting the refractive index. Rapidity and simplicity of measurement, the relatively inexpensive equipment required, the small sample size, and the

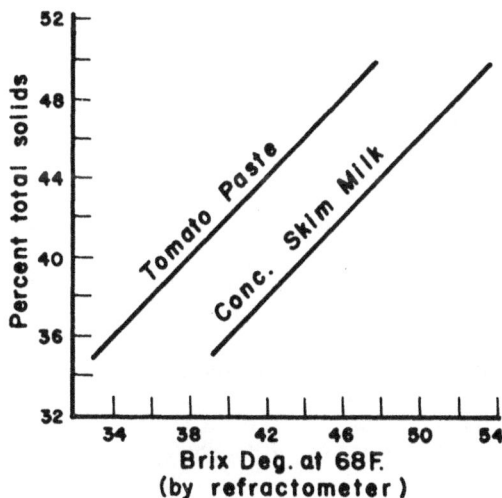

Courtesy of the American Optical Co.

FIG. 43. RELATION OF REFRACTOMETER DATA (IN BRIX DEGREES) TO TOTAL SOLIDS OF TOMATO PASTE AND CONDENSED SKIM MILK

nondestructive nature of the test are factors contributing to its popularity. Refractometry can be adapted to continuous in-line measurements (Maley 1963).

BIBLIOGRAPHY

ANON, 1960. Official Methods of Analysis of the Association of Official Agricultural Chemists, 9th Edition. Association of Official Agricultural Chemists, Washington, D. C.

ANON. 1962. Varian PA-7 Process Analyzer. Varian Associates, Palo Alto, Calif.

BENNETT, O. F. 1964. Water determination by gas chromotography. Anal. Chem. 36, 684.

BLISH, M. J., and HITES, B. D. 1950. A rapid and simple carbide method for estimating moisture in flour. Cereal Chem. 7, 99–107.

BLOCH, F., HANSEN, W. W., and PACKARD, M. 1946. The nuclear induction experiment. Physical Review 70, 474–485.

CARLSTROM, A. A., SPENCER, C. F., and JOHNSON, J. F. 1960. Determination of trace water in butane by gas chromatography. Anal. Chem. 32, 1056.

CHRISTENSEN, C. M., and LINKO, P. 1963. Moisture contents of hard red winter wheat as determined by meters and by oven drying, and influence of small differences in moisture content upon subsequent deterioration of the grain in storage. Cereal Chem. 40, 129–137.

COLE, L. G., CZUHA, M., MOSELY, W., and SAWYER, D. T. 1959. Continuous coulometric determination of parts per million of water in organic liquids. Anal. Chem. 31, 2048–2050.

CONWAY, T. F., COHEE, R. F., and SMITH, R. J. 1957. A new method of "moisture determination." Mfg. Confectioner 37, No. 5, 27, 29, 31, 32–34.

FISCHER, K. 1935. A new method for the analytical determination of the water content of liquids and solids. Angew. Chem. 48, 394–396.

GUILBOT, A. 1949. Considerations on the determination of water in biological substances. Comptes rend. acad. agr. France 36, 694–700.

HANCOCK, C. K., and BURDICK, R. L. 1956. Modified indirect conductivity method for determining water in cottonseed meal. J. Agr. Food Chem. 4, 800–802.

HLYNKA, I., and ROBINSON, A. D. 1954. Moisture and its measurement. In Storage of Cereal Grains and Their Products. Edited by J. A. Anderson and A. W. Alcock. American Association of Cereal Chemists, St. Paul, Minn.

KEIDEL, F. A. 1959. Determination of water by direct amperometric measurement. Anal. Chem. 31, 2043–2048.

KÖSTER, A. 1934. Moisture determination by drying techniques. Mühle 71, 1303–1310.

LINCOLN, H. W., DIRKS, B. M., and HARREL, C. G. 1954. A method for the rapid determination of moisture in doughs and breads. Cereal Chem. 31, 506–513.

MAKOWER, B. 1950. Determination of water in some dehydrated foods. Advances in Chem. Ser. No. 3, 37–54.

MALEY, L. E. 1963. Continuous measurement of dissolved solids in food process by critical angle refractometry. Food Technol. 17, 25–28, 30.

McNeil, H. C. 1912. A calcium carbide method for determining moisture. U. S. Dept. Agr. Circ. 97, 1–8.

Miller, B. S., and Kaslow, H. D. 1963. Determination of moisture by nuclear magnetic resonance and oven methods in wheat, flour, doughs, and dried fruits. Food Technol. 17, 650–653.

Mitchell, J., Jr., and Smith, D. M. 1948. Aquametry. Interscience Publishers, New York.

Neuss, J. D., O'Brien, M. G., and Friediani, H. A. 1951. Sodium tartrate dihydrate as a primary standard for Karl Fischer reagent. Anal. Chem. 23, 1332–1333.

Oxley, T. A., and Pixton, S. W. 1961. Sources of variation in moisture content results by oven methods. II. Possible causes of day to day variations. Milling 137, 242–250.

Palmer, K. J., and Elsken, R. H. 1956. Determination of water by nuclear magnetic resonance in hygroscopic materials containing soluble solids. J. Agr. Food Chem. 4, 165–167.

Penther, C. J., and Notter, L. J. 1964. Hydro-chromagraph, a process-type water analyzer combining chromatographic separation and electrolysis cell detection. Anal. Chem. 36, 283–286.

Purcell, E. M., Torrey, H. C., and Pound, R. V. 1946. Resonance absorption by nuclear magnetic moments in a solid. Physical Review 69, 37–38.

Rubin, H. 1958. New tool for moisture analysis. Nuclear magnetic resonance. Cereal Sci. Today 3, 240–243.

Rubin, H., and Swarbrick, R. E. 1961. Rapid fluorine analysis by wideline nuclear magnetic resonance. Anal. Chem. 33, 217–220.

Salwin, H. 1963. Moisture levels required for stability in dehydrated foods. Food Technol. 17, 1114–1123.

Schäfer, W., and Seibel, W. 1957. Problems and investigations in determining the moisture content of grain. Mühle 94, 663–670.

Schwecke, W. M., and Nelson, J. H. 1964. Determination of moisture in foods by gas chromatography. Anal. Chem. 36, 689–690.

Seaman, W., McComas, W. H., Jr., and Allen, G. A. 1949. Determination of water by Karl Fischer reagent. Anal. Chem. 21, 510–512.

Seibel, W., and Bolling, H. 1958. Automatic moisture determination with Karl Fischer solution in grain productions. Ber. Getreidechemiker Tagung, June 10–12.

Shaw, T. M., and Elsken, R. H. 1950. Nuclear magnetic resonance absorption in hygroscopic materials. J. Chem. Phys. 18, 1113–1114.

Shaw, T. M., and Elsken, R. H. 1956. Determination of water by nuclear magnetic absorption in potato and apple tissue. J. Agr. Food Chem. 4, 162–164.

Shaw, T. M., Elsken, R. H., and Kunsman, C. H. 1953. Moisture determination of foods by hydrogen nuclei magnetic resonance. J. Assoc. Offic. Agr. Chemists 36, 1070–1076.

Sloman, K. G., Borker, E., and Reussner, M. D. 1954. Determination of moisture in chocolate. J. Agr. Food Chem. 2, 1239.

Smith, D. M., Bryant, W. M. D., and Mitchell, J., Jr. 1939. Analytical procedures employing Karl Fischer reagent. I. Nature of the regeant. J. Am. Chem. Soc. 61, 2407–2412.

STEIN, H. H., and AMBROSE, J. M. 1963. Near-infrared method for determination of water in aluminum aspirin. Anal. Chem. 35, 550–552.

STITT, F. 1958. Moisture equilibrium and the determination of water content of dehydrated foods. In Fundamental Aspects of the Dehydration of Foodstuffs. Society of Chemical Industry, London.

WEISE, E. L., BURKE, R. W., and TAYLOR, J. K. 1964. Proceedings 1963 Symposium on Humidity and Moisture. Reinhold Publishing Co., New York.

ZELERY, L. 1960. Moisture measurement in the grain industry. Cereal Sci. Today 5, 130–134, 136.

CHAPTER 10

Measurement and Control of Humidity

MEASUREMENT OF HUMIDITY

A measurement of the amount of water vapor in a given volume of air or other gas is frequently necessary in food technology studies. Humidity affects the efficiency of many processing steps as well as the storage stability of most foods.

If the partial pressure of water vapor in a container of known volume can be measured, its quantity can be calculated. The vapor pressure of water at a given temperature is the pressure exerted by water vapor which is in contact with the liquid substance at that temperature. It can be measured directly by manometric techniques and exhaustive tables of such experimental data are available. The amount of water vapor which can be contained in a given space is controlled by the temperature and pressure. At ordinary pressures, the amount is independent of the kinds and amounts of other gases present for all practical purposes. At higher pressures, say about ten atmospheres and above, the water vapor pressure becomes significantly affected by the total pressure of the system in accordance with the relationship:

$$\frac{p}{p_0} = e^{\left(\frac{P}{D}\frac{M}{RT}\right)}$$

where p = vapor pressure of the liquid at total pressure P, p_0 = vapor pressure of the liquid in equilibrium with its vapor in the absence of other gases, D = density of the liquid and the other terms have the usual meaning.

Absolute humidity is the mass of water vapor present per unit of space. It is usually expressed in grams per cubic meter and can be computed by using the equation

$$a_h = 217\frac{e}{T_a}$$

where T_a is the absolute temperature in degrees Centigrade. Absolute humidity is equivalent to vapor density.

Specific humidity is the weight of water per unit weight of gas. It can be calculated according to the equation

$$\text{Specific humidity} = \frac{0.622\,P_w}{P_t - P_w}$$

where P_w is the partial pressure of the water vapor and P_t is the total pressure inclusive of the partial pressure of the water vapor. Absolute humidity and specific humidity are less useful measurements for most purposes than relative humidity, which is the partial pressure of water divided by the saturated vapor pressure of the liquid at the same temperature. It is expressed as a percentage.

If two or more solids or liquids containing water are present together in a closed system, a transfer of water vapor will occur until the moisture contents of all of the system's components are in equilibrium with the same relative humidity. The rate at which equilibrium is approached in such systems depends on many factors, among which are temperature, particle size (surface area) of the solids, gas currents inside the container, separation of the components, and spread between the initial equilibrium relative humidities.

The most reliable means for determining the quantity of water vapor in a given volume of air is to separate it out and weigh it directly. Although it is possible to collect water vapor as ice on surfaces held at temperatures so low that the vapor pressure of the water is negligible, it is much simpler to remove it by passing the gas through efficient desiccants such as phosphorus pentoxide. If sufficient contact between the gas and the desiccant is obtained, all water within the usual limits of accuracy of weight measurement will be removed. When temperature and the volume of the gas sample are known, the difference in weight of the desiccant after the experiment will be all that is required to calculate the absolute humidity. Practical difficulties of instrumentation and procedure are encountered in this technique.

Wexler and Hyland (1963) describe such a gravimetric hygrometer. This instrument removes, for weighing, the moisture from a sample of gas flowing at a constant rate through a series of absorption tubes and into an evacuated cylinder of known volume. The absorption tubes are filled with magnesium perchlorate or phosphorus pentoxide. Gain in weight, equal to the amount of water taken up, is determined by precise, but conventional, methods. Flow of the gas sample through the absorption tubes is maintained at the desired rate by drawing it into stainless-steel cylinders of known volume. Mass of the sample is calculated from the number of times the cylinders are filled, the cylinder volumes, and the pressure and temperature measured in each cylinder when filled.

It has been estimated that humidity can be determined with a maximum error of 12 parts in 10,000 (0.12%) by the gravimetric hygrometer. This high degree of accuracy has led the U. S. Bureau of Standards to adopt the device as a humidity standard for calibration of other instruments, but the precision required in measuring the weights and volumes would seem to rule out this procedure for ordinary work. For most purposes, there are methods of sufficient accuracy which use less elaborate instrumentation. The psychrometer is especially prominent in meteorological studies but has also been much used in food studies.

A psychrometer consists essentially of two thermometers, one of which is operated in the normal manner, while the other is mounted so that its bulb is constantly in contact with a film of evaporating water (see Fig. 44). The reading of the wet-bulb thermometer is thus depressed below

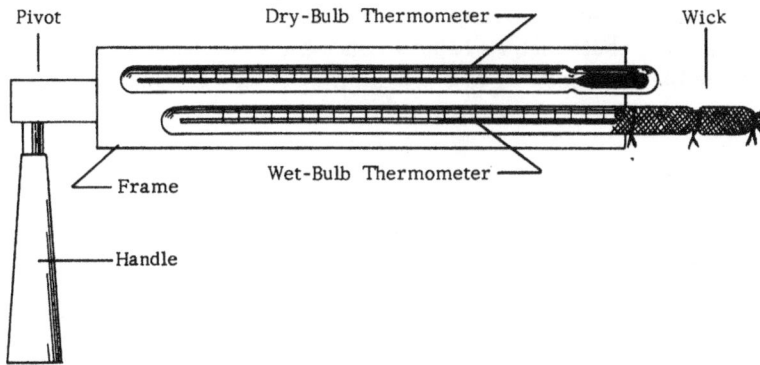

Courtesy of Hygrodynamics, Inc.

FIG. 44.   SLING PSYCHROMETER

the reading of the dry-bulb thermometer at all relative humidities below 100%. The extent of this depression is dependent upon the rate of evaporation of water from the film surrounding the wet-bulb thermometer and this, in turn, depends upon the relative humidity of the ambient space. The extent of the temperature depression due to the evaporating water and the reading of the dry bulb thermometer are the only data required to determine the relative humidity by consulting a psychrometric chart.

Usually, required contact with water is obtained by inserting the lower part of the thermometer into a cloth tube which extends into a container of distilled water. The tube or wick is kept wet by capillary action. The water reservoir must be maintained at the temperature of the dry bulb thermometer, i.e., at the temperature of the gas being measured. Use of colder or warmer water, even though the difference is only a fraction of a degree, will introduce appreciable error into the determination. The wick should be changed frequently, since accumulation of soil will affect its performance. A repeatedly washed wick is better than a new wick since the latter contains a natural oil which is gradually eliminated by washing.

Means for securing a rapid flow of air over the sensing regions of the psychrometer must also be provided. It has been found that a stream of air having a velocity of about 600 f. p. m. causes the water held by the wick to evaporate at a rate sufficient to yield an accurate and stable wet-bulb reading. The sling psychrometer is manually whirled about by the operator in order to secure the desired air flow. In fixed permanent installations, fans are used to move a stream of gas over the thermometer bulbs. If the fan is mounted on a motor shaft, it should be placed so that the air is sucked over the wick rather than blown over it, since heat radiated from the motor casing will otherwise affect the temperature of the air.

Although the usual practice is to determine the relative humidity from a psychrometric chart such as Table 15 once the wet- and dry-bulb readings are available, it is possible to calculate it after first calculating the vapor pressure e according to the equation:

$$e = e_s - 0.000367 \, p_a \, (T - T_w)\left(1 + \frac{T_w - 32}{1571}\right)$$

where T and T_w are the dry- and wet-bulb temperatures in degrees Fahrenheit, and e_s is the saturation vapor pressure in millibars at T_w. When the vapor pressure e is known, the relative humidity can be calculated using the relationship:

$$\text{Per cent relative humidity} = 100\frac{e}{e_s}.$$

Because of the many possible observational errors which are possible, the standard psychrometer must be used with great caution. Even so, it cannot be relied upon for measurements of high precision. The two thermometers double the chance of misreading, and, at low temperatures, misreading by a few tenths of a degree can lead to gross errors. Taking a reading before the wet-bulb thermometer has reached its lowest point is a common mistake. Insufficient air flow, dirty wicks or wicks that are too thick, and impure water lead to inaccurate observations.

When approximate indications of relative humidity are acceptable, hygrometers based on measurements of the contraction or expansion of hair or certain other animal tissues can be used. They are relatively sturdy and inexpensive, function over a wide range, and give direct readings. The principle is simple. Strands of hair are brought from a fixed point around a spring-loaded cylinder or lever which has an indicator attached to it. As the humidity changes, the hair lengthens or contracts, exerting greater or lesser torque on the cylinder. These devices are subject to appreciable error. The hair expands with increasing temperature, and its response to changes in humidity is not strictly linear and is very slow, the lag increasing with decreasing temperature.

TABLE 15 PER CENT RELATIVE HUMIDITY FROM WET- AND DRY-BULB READINGS

Difference Between Readings of Wet and Dry Bulbs in Degrees Fahrenheit

Dry Bulb, Deg. F.	1	2	3	4	5	6	7	8	9	10	11	12	13	14	15	16	17	18	19	20	21	22	23	24	25	26	27	28	29	30	Dry Bulb, Deg. F.
40	92	83	75	68	60	52	45	37	29	22	15	7	0																		40
41	92	84	76	69	61	54	46	39	31	24	17	10	3																		41
42	92	84	76	69	62	54	47	40	33	26	19	12	5																		42
43	92	85	77	70	62	55	48	42	35	28	21	14	8	1																	43
44	93	85	77	70	63	55	48	42	35	28	23	16	10	4																	44
45	93	85	78	71	63	56	49	44	37	30	25	18	12	6																	45
46	93	86	78	71	64	57	51	44	38	32	26	20	14	8	2																46
47	93	86	79	72	65	58	52	45	39	34	28	22	16	10	5	1															47
48	93	86	79	72	66	58	53	46	40	35	29	23	18	12	7	3															48
49	93	86	79	73	66	59	54	47	41	36	32	25	19	14	9	5															49
50	93	87	80	73	67	60	54	49	43	36	34	27	21	16	10	7	0														50
51	94	87	80	74	67	61	55	50	45	38	35	28	23	17	12	9	2														51
52	94	87	80	74	68	61	56	51	46	39	37	29	24	19	14	10	4														52
53	94	88	81	75	69	62	57	52	48	40	38	31	26	20	16	12	6	1													53
54	94	88	81	75	69	63	58	53	50	41	40	32	27	22	17	14	8	3													54
55	94	88	82	75	70	63	59	54	51	42	41	33	28	23	19	16	9	5													55
56	94	88	82	76	70	64	59	55	52	43	42	34	30	25	20	17	11	5													56
57	94	88	82	76	71	65	60	55	53	45	43	35	31	26	22	18	13	7	0												57
58	94	89	82	77	71	65	61	56	53	46	45	37	32	27	23	20	14	8	2												58
59	94	89	83	77	72	66	61	57	54	47	47	38	33	29	24	21	16	10	4												59
60	95	89	83	78	72	66	62	58	54	48	48	39	34	30	26	22	17	11	6	0											60
61	95	89	84	78	73	67	62	58	55	49	48	40	35	31	27	24	18	12	7	1	1										61
62	95	89	84	78	73	67	63	59	56	50	50	41	36	32	28	25	20	14	9	3	3										62
63	95	90	84	79	73	68	63	60	56	50	51	42	37	34	30	26	21	16	10	5	4	1									63
64	95	90	84	79	74	68	64	60	57	51	52	43	38	35	31	27	22	17	11	7	6	2									64
65	95	90	84	79	74	69	65	61	58	52	53	44	39	36	32	29	24	18	13	8	7	4	0								65
66	95	90	85	80	75	69	65	61	58	53	54	45	40	37	33	30	25	20	15	10	9	5	2								66
67	95	90	85	80	75	70	66	62	59	54	55	46	42	38	34	31	26	21	16	11	10	7	3	0							67
68	96	90	85	80	75	70	66	62	59	54	56	47	43	39	35	32	27	22	17	12	12	8	5	2							68
69	96	91	85	80	76	71	67	63	60	55	57	48	44	40	36	33	28	23	19	14	13	10	6	3							69
70	96	91	86	81	76	71	67	64	61	56	57	48	45	41	37	34	29	24	20	15	14	11	8	5							70
71	96	91	86	81	76	71	68	65	61	57	59	49	46	42	38	35	30	25	22	16	15	12	9	6	1						71
72	96	91	86	81	77	72	68	65	61	57	59	50	47	43	39	36	31	27	23	18	17	13	10	7	3						72
73	96	91	86	82	77	72	69	66	62	58	60	50	47	44	40	37	32	28	24	19	18	15	12	9	4						73
74	96	91	87	82	77	72	69	66	62	58	61	51	48	45	41	38	33	29	25	20	19	16	13	10	6	1					74
75	96	91	87	82	78	73	69	67	63	59	62	51	48	46	42	39	34	30	27	21	20	17	14	11	7	3					75
76	96	91	87	82	78	73	70	67	63	59	63	52	49	47	43	40	35	31	28	22	21	18	15	12	8	4					76
77	96	92	87	83	78	74	70	68	64	60	64	53	50	48	44	41	36	32	29	23	22	19	16	13	9	5	1				77
78	96	92	87	83	79	74	71	68	64	61	65	53	50	49	46	42	37	34	30	24	23	20	17	14	11	7	3				78
79	96	92	88	83	79	74	71	69	65	61	66	54	51	50	46	43	38	35	31	25	24	21	18	15	12	8	4	1			79
80	96	92	88	83	79	75	71	69	65	62	66	55	52	51	47	44	40	37	32	26	25	22	19	16	13	9	5	3	5	3	80
82	96	92	88	84	80	75	72	70	66	63	67	56	53	52	48	44	40	38	33	27	26	23	20	17	14	10	6	4	7	5	82
84	96	92	88	84	80	76	73	70	66	64	68	57	54	53	49	46	42	39	35	28	27	24	21	18	15	11	8	5	9	7	84
86	96	93	88	84	81	76	73	71	67	64	69	57	54	54	50	47	43	40	35	29	28	25	22	19	16	12	9	6	11	9	86
88	96	93	89	85	81	77	74	72	68	65	70	58	55	55	51	48	44	41	37	30	29	26	24	20	17	14	10	7	13	11	88
90	96	93	89	85	81	77	74	72	68	66	71	59	56	56	52	49	45	42	39	32	32	28	26	21	19	16	12	9	15	13	90
92	96	93	89	85	82	77	75	73	69	66	71	59	56	56	53	50	46	43	40	33	33	30	27	23	21	18	14	10	16	15	92
94	96	93	89	85	82	78	76	73	69	67	72	60	57	57	54	50	47	44	41	35	34	32	28	25	22	20	16	12	18	16	94
96	96	93	89	86	82	79	76	73	70	68	73	61	58	58	54	51	48	45	42	36	35	33	30	26	24	22	18	14	20	18	96
98	96	93	89	86	82	79	76	74	70	68	73	61	58	58	55	52	49	46	43	37	36	34	31	28	25	23	19	15	21	19	98
100	96	93	89	86	83	80	77	74	70	68	74	62	59	59	56	52	49	46	44	38	37	35	34	33	30	24	21	17	22	21	100

Another approach to hygrometer construction utilizes a diaphragm of hygroscopic animal membrane clamped between two aluminum retaining rings, forming a flat truncated cone surface. Changes in the humidity of the air surrounding the assembly cause directly proportional changes in dimensions of the diaphragm. Movements of its apex are communicated to the indicating or recording mechanism by a multiplying linkage connected to a small ring affixed to the center of the diaphragm.

When animal tissues are used as the basis of hygrometer movements, they must be properly aged and calibrated to minimize changes in usage.

Electrical Hygrometers

Hygrometers which depend for their action on the changes in electrical parameters of certain substances accompanying variations in relative humidity are coming into wide use in all branches of science and industry. The basic electrical hygrometer consists of an insulator supporting two metal electrodes joined by a film of hygroscopic salt which absorbs moisture in amounts dependent upon temperature and relative humidity of the air. The quantity of water absorbed governs the resistance of the system. These instruments are calibrated by plotting the resistance against relative humidity at several fixed temperatures.

A typical sensor has a lithium chloride hygroscopic film coating a bifilar-wound, noble-metal resistance element (Quinn 1963). The resistance of the hygroscopic film responds to small changes in the vapor pressure. The wire electrodes may be wound on a textile sleeve impregnated with the hygroscopic substance. Honeywell's (Anon. 1962) humidity sensor is composed of two intermeshing gold grids stamped on plastic and is coated with a complex film containing hygroscopic salts. As relative humidity rises, the film becomes more conductive and electrical resistance between the grids is lowered. The variation of electrical resistance is interpreted by a recording or controlling unit. This sensor is said to be capable of detecting a small fraction of one per cent change in relative humidity.

Lithium chloride is the hygroscopic substance most frequently used in the sensors of electronic hygrometers. It has two characteristics which make it particularly suitable for this usage. It has a strong affinity for water vapor and it has an inherent ability to maintain itself at a constant value just above 11 per cent relative humidity when it is in a moist atmosphere and heated by an electric current passing through it. At values of 11 per cent relative humidity and below, the lithium chloride dries to a crystalline solid and becomes practically non-conductive.

The hygroscopic characteristics of different salts can be used as a direct indication of the relative humidities of atmospheres. Pouncey and Summers (1939) placed crystals of salts which deliquesce at known relative humidities in small closed holes bored in the sample, and observed the condition of the crystals after several hours of incubation. Microscopic observations indicated that small drops of water had formed around some of the crystals and they were beginning to dissolve. The equilibrium relative humidity of the sample was regarded as being equal to the highest equilibrium relative humidity of the salt that was dissolving.

A somewhat similar technique was developed by Kvaale and Dalhoff (1963) who used strips of filter paper impregnated with $ZnSO_4$, Na_3C_6-H_5O_7 (sodium citrate), $BaCl_2$, $Na_2C_4H_4O_6$, KNO_3, $NaBrO_3$, Na_2SO_4, Pb-$(NO_3)_2$, or Na_2HPO_4 for measuring relative humidity of pre-packaged meat products. When the strips are placed in a closed container with the sample and the container is maintained at 68°F. for 20 hr., the wet or dry state of the papers can be used as a direct indication of the equilibrium relative humidity of the sample. These papers are useful in the range of 90 to 100% relative humidity.

It is possible to determine humidity by absorption spectroscopy, although this technique is not much used because of interference by other substances at the useful wavelengths, and other problems. Water vapor is transparent to visible frequencies, but it absorbs strongly in the infrared. Several strong lines in the vicinity of 33.0μ wavelength are suitable for measurement. This method is probably more applicable to meteorology than to the disciplines directly related to food technology.

Radio refractometry has been applied to humidity measurements. Water vapor has a pronounced effect on the speed at which microwave radiation travels through it. Vetter and Thompson (1962) describe a microwave refractometer which can be used to determine absolute humidity with an accuracy of better than ±0.2 gm. of water per cubic meter of air when corrections for sample temperature and barometric pressure are applied. This is equivalent to about ± two per cent accuracy in measurement of absolute humidity at average sea level conditions and perhaps ± 15% under arid conditions.

The radio refractometer used by McGavin and Vetter depends upon the relationship between the resonance frequencies of a microwave cavity and the contents of the cavity. The microwave refractive indices of a sample cavity and a reference cavity are compared electronically.

Wildhack (1950) suggested a pneumatic bridge arrangement for determining humidity and this principle has recently been utilized in the design of working instruments (Wildhack et al. 1963; and Greenspan 1963). The pneumatic bridge is in some ways analogous to the common electrical (Wheatstone) bridge used for comparing parameters of electric current. The diagram in Fig. 45 illustrates the similarities. Nozzles, Y_1, X_2, and Y_2, are analogous to resistances R_Y, R_A, and R_B while nozzle X_1 and the

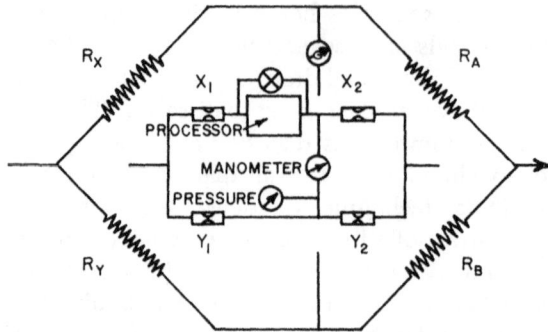

From Wildhack et al. (1963) and Greenspan (1963)

FIG. 45. PNEUMATIC BRIDGE METHOD FOR DETERMINING HUMIDITY OF GASES IN COMPARISON WITH WHEATSTONE BRIDGE

Refer to text for explanation of symbols.

"Processor" are together analogous to the unknown resistance in the electrical bridge. In the pneumatic bridge, the gas is in "critical flow," that is, the rate of gas flow through each bridge constriction is determined solely by the pressure and temperature upstream of the constriction and is independent of the downstream pressure. This condition occurs when the ratio of the throat pressure to the upstream pressure falls below a critical value, about 0.5 for air. In the bridge, a pressure differential between the midpoints of the flow paths is observed when gas in one branch has been changed in volume by any process taking place just upstream of the point of measurement in that branch. A steady flow of the gas sample is split at the input into two parallel paths, a reference branch and a test branch, each incorporating two constrictions or nozzles. A desiccator (or a saturator) is inserted between the nozzles of the test branch. The subtraction (or addition) of water vapor pressure changes the total pressure. This creates an imbalance between pressures at the entrances to the two downstream nozzles; measurement of one pressure and the differential can then be used to calculate the humidity.

Humidity can also be determined by gas chromatography. This is a rapid, practical technique which is finding wider use each day. Analysis of headspace gases from containers and the determination of moisture in grain are two of the earlier applications. The principle is described in the preceding chapter on moisture analysis.

Dew point is the temperature at which the partial pressure of water in a gas is equal to the saturated vapor pressure of water. In other words, a gas at its dew point would have a relative humidity of 100%; it would be saturated. A further release of moisture into the space, or a decrease of temperature, would cause condensation, or "dew," to form. The dew point of a gas is thus a measure of the moisture in the gas in terms of the partial pressure of the water vapor. It is independent of the temperature, and, for almost all purposes, independent of the type of gas. Dew point or frost point hygrometers measure the temperature at which dew or frost is deposited on a cooled surface, usually a highly polished mirror. The dew point is an important datum in many freezing and drying applications, since the condensation of droplets of moisture on the surface of a food can cause damage far out of proportion to the percentage moisture increase relative to the whole food.

CONTROL OF HUMIDITY

Enclosed places in which the relative humidity can be controlled are important tools in many industrial processes and laboratory techniques. The humidified fermentation rooms and proof "boxes" found in most bakeries are well-known industrial applications. In these cases an atmosphere of high humidity is required in order to prevent the drying out or crusting of the surfaces of dough pieces during the fermentation or proofing periods. On the other hand, if the humidity is allowed to rise to near the dew point, there will exist the danger of water precipitating on the dough with the resultant development of a sticky and spotted surface. The solution lies in the careful control of the relative humidity within a narrow, high range. In most of these installations, only a limited choice of dry-bulb temperatures is available to the operator. The humidity is maintained by the intermittent injection of low pressure steam or atomized water actuated by a fall in wet-bulb temperature. Favorable conditions for fermentation rooms are a temperature of 80°F. and a relative humidity of 75%; for pan-proofing temperatures of 100° to 120°F. and a relative humidity of about 82% are desired.

Careful control of the ambient relative humidity is desirable in the successful production of alimentary pastes and such confections as marshmallow as well as in dehydration operations of all kinds. Efficiency and product quality could be improved in many other food processing operations if adequate humidity control were employed. Industries other than food manufacturing have made greater use of humidity control.

Laboratory humidistats sometimes make use of mechanical and electronic methods for control of water vapor, especially in the larger size units (see Fig. 46), but more frequently they are the static types in which the atmosphere within the enclosure, such as a desiccator jar, is controlled

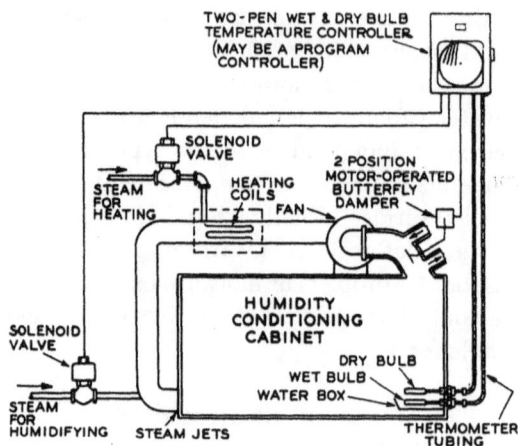

Courtesy of the Bristol Co.

Fig. 46. Typical Installation Diagram for a Humidity
Conditioning Cabinet

by its contact with a saturated solution of a suitable salt or with a sulfuric
acid solution.   The entire container is placed in a constant temperature
cabinet.   Pressures are almost always uncontrolled.   Provision for air circulation within the jar is absent and attainment of equilibrium between the
solution, atmosphere, and sample may take days or even weeks.   In the interim, chemical, enzymatic, or microbiological changes may modify the
sample sufficiently to affect its equilibrium relative humidity.

Tables 16 and 17 contain data on the relative humidities of atmospheres
in equilibrium with various saturated salt solutions and sulfuric acid solutions.   During the period prior to the establishment of an equilibrium in
these atmospheres, the solutions will gain or lose moisture.   As long as the
salt solutions are in contact with an excess of the solid chemical, their
equilibrium relative humidities will not be altered, but small changes in
the humidity of the atmosphere in equilibrium with a sulfuric acid solution can be expected as the solution gains or loses water.   These changes
are of no practical significance in most experiments.

Wilson (1921) lists the following advantages of sulfuric acid solutions
for control of humidity in the laboratory: (1) homogeneous solutions
varying from 0 to 100% water can be obtained; (2) the vapor pressures of
these solutions have been as accurately determined as those of any other
concentrated solution; (3) the composition and hence the vapor pressure
of a solution can be quickly and accurately determined by measuring its
density, which varies greatly with changes in composition; (4) their relative vapor pressure—percentage of that of water at the same temperature—

TABLE 16

VAPOR PRESSURES ABOVE SULFURIC ACID SOLUTIONS AT FOUR TEMPERATURES[1]

Concentration of Sulfuric Acid Per Cent	Relative Vapor Pressure Values[2] at			
	32°F.	77°F.	122°F.	167°F.
0	100.0	100.0	100.0	100.0
5	98.4	98.5	98.5	98.6
10	95.9	96.1	96.3	96.6
15	92.4	92.9	93.4	93.8
20	87.8	88.5	89.3	90.0
25	81.7	82.9	84.0	85.0
30	73.8	75.6	77.2	78.6
35	64.6	66.8	68.9	70.8
40	54.2	56.8	59.3	61.6
45	44.0	46.8	49.5	52.0
50	33.6	36.8	39.9	42.8
55	23.5	26.8	30.0	33.0
60	14.6	17.2	20.0	22.8
65	7.8	9.8	12.0	14.2
70	3.9	5.2	6.7	8.3
75	1.6	2.3	3.2	4.4
80	0.5	0.8	1.2	1.8

[1] From Wilson (1921) who calculated the values from data of others.
[2] The relative vapor pressures are given as percentages of that of pure water at the same temperature and therefore correspond to relative humidity of the air.

TABLE 17

RELATIVE HUMIDITIES AND VAPOR PRESSURES OVER SATURATED-SALT SOLUTIONS[1]

Chemical	At 73°F.		At 86°F.		At 100°F.	
	V.P.	R.H., %	V.P.	R.H., %	V.P.	R.H., %
Water	20.79	...	31.82	...	49.10	...
Ammonium monophosphate	19.32	92.9	29.27	92.0	44.72	91.1
Potassium chromate	17.98	86.5	27.47	86.3	42.05	85.6
Ammonium sulfate	16.65	80.1	25.32	79.6	38.85	79.1
Sodium chloride	15.69	75.5	23.94	75.2	36.88	75.1
Sodium acetate	15.56	74.8	22.72	71.4	33.23	67.7
Sodium nitrite	13.47	64.8	20.15	63.3	30.33	61.8
Sodium bromide	12.16	58.5	17.93	56.3	26.37	53.7
Sodium dichromate	11.24	54.1	16.56	52.0	24.56	50.0
Magnesium nitrate	11.13	53.5	16.34	51.4	24.06	49.0
Potassium nitrite	10.10	48.6	15.03	47.2	22.53	45.9
Calcium nitrate	10.77	51.8	14.83	46.6	19.11	38.9
Potassium thiocyanate	9.68	46.6	13.92	43.7	20.17	41.1
Potassium carbonate	9.12	43.9	13.85	43.5	21.32	43.4
Chromium trioxide	8.16	39.2	12.74	40.0	19.73	40.2
Magnesium chloride	6.83	32.9	10.32	32.4	15.65	31.9
Potassium acetate	4.76	22.9	6.99	22.0	10.00	20.4
Lithium chloride	2.31	11.1	3.56	11.2	5.47	11.1

[1] From Wink and Sears (1950).

varies but little with wide changes in temperature; (5) they come to equilibrium rapidly with the surrounding atmosphere; (6) the sulfuric acid itself exerts no appreciable vapor pressure; and (7) material of adequate purity is cheap and easily obtainable. To these might be added the advantage that sulfuric acid undergoes no changes of importance during storage and use (i.e., no hydrolysis, polymerization, etc.).

When a stream of air of controlled humidity is required, it can be obtained by mixing appropriate quantities of two streams of air, one of which has been thoroughly dried and another of which has been saturated as by bubbling through water. By varying the relative quantities of the two streams it is possible to obtain any desired relative humidity. Another, less cumbersome, method involves bubbling the stream of air through an aqueous sulfuric acid solution of the appropriate concentration.

BIBLIOGRAPHY

ANON. 1962. Humidity—How it Affects your Industry and What You Can Do About It. Honeywell Corp., Minneapolis, Minn.

ARNOLD, J. H. 1933A. The theory of the psychrometer. 1. The mechanism of evaporation. Physics 4, 255–262.

ARNOLD, J. H. 1933B. The theory of the psychrometer. II. The effect of velocity. Physics 4, 334–340.

ATKINS, R. M. 1964. Wet-dry bulb thermistor hygrometer with digital indication. Instr. and Control Systems 73, No. 4, 111–114.

FLANIGAN, F. M. 1960. Comparison of the accuracy of the humidity measuring instruments. ASHRAE J. 2, 56–59.

FOSKETT, L. W., FOSTER, N. B., THICKSTUN, W. R., and WOOD, R. C. 1953. Infrared absorption hygrometer. Monthly Weather Rev. 81, 267–277.

GATES, D. M., VETTER, M. J., and THOMPSON, M. C. JR. 1963. The measurement of moisture boundary layers and leaf transpiration with a microwave refractometer. Nature 197, 1070–1072.

GRALENSKI, F. M. 1964. Automatic dewpoint hygrometer. Instr. and Control Systems 37, No. 5, 124–125.

GREENEWALT, C. H. 1926. Absorption of water vapor by sulfuric acid solutions. Ind. Eng. Chem 18, 1291–1295.

GREENSPAN, L. 1964. A pneumatic bridge hygrometer for use as a working humidity standard. Proc. 1963 Symposium on Humidity and Moisture. Reinhold Publishing Corp., New York.

HUGHES, F. J., TESSEN B. M., and KOCH, R. B. 1965. Characteristics of humidity elements used in flour moisture measurements. Cereal Sci. Today 10, 6, 8, 12.

KVAALE, O., and DALHOFF, E. 1963. Determination of the equilibrium relative humidity of foods. Food Technol. 17, 659–661.

LANDROCK, A. H., and PROCTOR, B. E. 1951. A new graphical interpolation method for obtaining humidity equilibrium data, with special reference to its role in food packaging studies. Food Technol. 5, 332–337.

Morris, V. B., Jr., and Sobel, F. 1954. Some experiments on the speed of response of the electric hygrometer. Bull. Am. Meteorol. Soc. 35, 226–229.

Pouncey, A. E., and Summers, B. C. L. 1939. The micromeasurement of relative humidity for the control of osmophilic yeasts in confectionery products. J. Soc. Chem. Ind. 58, 162–170.

Quinn, F. C. 1963. Equilibrium hygrometry. Instr. and Control Systems 36, No. 7, 113–114.

Schaffer, W. 1946. A simple theory of the electric hygrometer. Bull. Am. Meteorol. Soc. 27, 147–151.

Vetter, M. J., and Thompson, M. C., Jr. 1962. An absolute microwave refractometer. Rev. Sci. Instr. 33, 656–660.

Weise, E. L., Burke, R. W., and Taylor, J. K. 1963. Determining moisture in grain by gas chromatography. Nat. Bur. Standards (U. S.) Tech. News Bull. 47, 116–117.

Wexler, A., and Hyland, R. 1963. Gravimetric hygrometer for precise measurement of humidity. Natl. Bur. Standards (U. S.) Tech. News Bull. 47, 122–123.

Wildhack, W. A. 1950. A versatile pneumatic instrument based on critical flow. Rev. Sci. Instr. 21, 25–30.

Wildhack, W. A., Perls, T. A., Kissinger, G. W., and Hayes, J. W. 1963. NBS Hygrometers make precise humidity measurements by means of critical flow through pneumatic bridge. Natl. Bur. Standards (U. S.) Tech. News Bull. 47, 118–121.

Wilson, R. E. 1921. Humidity control by means of sulfuric acid solutions, with critical compilation of vapor pressure data. J. Ind. Eng. Chem. 13, 326–331.

Wink, W. A., and Sears, G. R. 1950. Instrumentation studies. LVII. Equilibrium relative humidities above saturated salt solutions at various temperatures. Tappi 33, No. 9, 96A–99A.

CHAPTER 11

Water Distribution Systems

INTRODUCTION

Although the actual construction of a distribution system is best left to specialists in that field, a knowledge of the principles involved in selecting the necessary equipment and designing the system will enable the food scientist to make more valid judgments of the adequacy and cost of proposed construction. Space limitations necessitate a cursory discussion of the pertinent topics in this chapter. Complete treatments of the subject can be found in hydraulics texts and in the thorough discussions of Skeat (1958) and Babbitt et al. (1962).

A water distribution system for a food plant will ordinarily include pipes, valves, pumps, and tanks, and may include storage or equalizing reservoirs, wells, fire fighting equipment, etc. The adequacy of any distribution facility is determined by the availability of a sufficient supply of water of suitable quality at the place and time it is needed. In the design of systems, this is usually estimated by calculating the pressures which will exist at points of withdrawal under presumed conditions of operation.

The initial source of pressure which actuates the flow of water through the pipes is either gravity or pumps. The energy for gravity flow is supplied by transferring the water to reservoirs or tanks which are elevated with respect to the rest of the system. Usually this is done by pumps although in unusual cases it is possible to direct water from a natural source into an elevated reservoir. If water from a municipal plant is used, its pressure may have to be supplemented by the user.

Factors causing loss of pressure are friction, removal of water from the system, and fittings (turbulence). Often other factors are ignored and loss of head is regarded as due only to friction in straight pipe. The relation of pipe diameters to head loss due to friction, and rate of flow is given by the formulas (Williams and Hazen 1920):

$$V = 0.0131\ CH^{0.54}D^{0.632}$$

$$Q = 0.0103\ CH^{0.54}D^{2.63}$$

where

$V =$ velocity in feet per second
$Q =$ rate of flow in cubic feet per second
$H =$ head loss due to friction, feet per
1,000 ft. of straight pipe.

D = diameter of flow, feet

C = the friction coefficient which is empirically determined

The values for C depend upon the texture of the internal surface of the pipe. Very complete tables of these constants have been published. Some typical Hazen and Williams friction coefficients for circular pipes running full are:

New cast-iron pipe, pit cast.............................120–130
New cast-iron pipe, centrifugally cast.....................125–135
New tar-coated pipes, smaller than 16 in. diam............125
Transite pipe, 6 in. diam.................................140+
Bitumastic enamel lining centrifugally applied............145–155
Ordinary tar-dipped cast iron, new.......................135
    Same, after five years average service.................120
    Same, after 15 years average service....................105
    Same, after 30 years average service.. 85

When hydraulic engineers speak of a "95" pipe or a "130" pipe, it is to these constants that they are referring.

Solution of Hazen and Williams' equations has been simplified by the publication of tables, charts, and nomographs. Direct computation requires the use of logarithmic tables, a slide rule, or a computer.

Where the pressure to the system is supplied by pumps, the size of the pipe as well as the nature of its internal surface will affect energy requirements. If the costs of the different sizes of pipe and the cost of pumping per unit of work done are known, the diameters of the distribution lines can be selected so as to minimize the amount of electricity or fuel consumed per year. Generally, the kind and capacity of the pumps needed in a plant system will not be much changed by the minor design variations resulting from adjustments in the dimensions of the distribution lines, although this may not be true for large municipal networks.

The preliminary estimates of pipe sizes needed to give required rates of flow in complex systems rely heavily on experience and trial-and-error computations. In computing head losses due to friction, the equivalent-pipe method is widely used. In this procedure, computations from the layout are greatly simplified by omitting and combining pipes, converting sizes, etc. to form series (end-to-end combinations of pipes having different diameters) and parallel (side-by-side combinations of pipe having the same diameter) compound pipes. One pipe combination is said to be equivalent to another when the losses of head for equal rates of flow in them are the same.

Babbitt *et al.* (1962) described a method for the solution of problems involving head losses and rates of flow in compound pipes. The first step requires the changing of all the pipes of the system to equivalent lengths of eight-inch diameter pipe with the constant C being 100. This operation is performed by substituting the appropriate data in the expression:

$$l_8 = \left(\frac{100}{C_n}\right)^{1.85}\left(\frac{8}{d_n}\right)^{4.87} l_n$$

where

l_8 = equivalent length in feet of eight-inch diameter pipe having $C = 100$

C_n = hydraulic coefficient for pipe having diameter d_n

l_n = length, in feet, of the pipe having diameter d_n

In the solution of a problem involving a parallel compound pipe, each pipe should be converted, for convenience, to an equivalent length of eight-inch pipe whose $C = 100$. If the total rate of flow through the system is known, the rate of flow through each pipe will be inversely proportional to its length to the 0.54 power, and the head loss through the system will be equal to the head loss through any one pipe.

EFFECT OF FITTINGS

Energy (head, pressure) is lost at the entrance and discharge points of a distribution system as well as at enlargements, contractions, valves, and other fittings. By experiment, it has been found that these losses are roughly proportional to the velocity of the fluid. In distribution systems which consist largely of straight pipes, these auxiliary losses are sometimes neglected in calculating the total head required and other factors.

The energy lost at entrances and contractions is in accordance with the formula:

$$h = k\frac{V^2}{2g}$$

where

V = velocity in the pipe past the fitting or entrance, in feet per second

g = acceleration of gravity, feet per second

h = head loss, feet

k = a coefficient which is related to V and to the ratio of pipe diameters at contractions. Tables of these coefficients have been published. Some typical values are:

Ratio of smaller to larger diameter—0.20 0.40 0.60 0.80
Value of k—0.46 0.41 0.28 0.10

Sudden enlargements give a head loss described by the empirical formula:

$$h_e = \frac{(V_a - V_b)^2}{2g} = \frac{V_a^2}{2g}\left[1 - \left(\frac{d_a}{d_b}\right)^2\right]^2$$

where V_a is the velocity in the smaller pipe, V_b the velocity in the larger pipe, d_a and d_b the respective diameters, and the other symbols have the same meaning as previously given.

Excess loss of head (i.e., over that in a straight pipe of the same length and diameter) in 90° bends can be obtained to a fair degree of accuracy from the equation $h = k(V_2/2g)$, where k is a coefficient related to the length of the bend. For 45° bends, three-fourths of these values, and for $22^1/_2°$ bends one-half of these values can be assumed.

Different types of valves cause different amounts of head loss in water passing through the fully opened fitting. The loss may also be related to the size of the valve and certain other features, depending on the circumstances and type of valve. The general expression $h = k/(V^2/2g)$ is applicable here. The value of k for globe valves from $^3/_4$ to 2 in. in diameter varies from 6 to 7.5. Constants for other types of fittings can be found in books of hydraulic tables.

HYDRAULIC GRADIENT

In designing rather large distribution systems, the concept of the hydraulic grade line or hydraulic gradient comes into play. The hydraulic grade line is a locus joining the points whose vertical distances from the centroid of the cross-section of a stream flowing in a closed channel are proportional to the pressures in the pipe at the point. On a map or diagram of a pipe-line, the hydraulic grade line can be regarded as a graph of the pressures existing along the pipe-line. The slope depends upon the frictional loss. When no flow is occurring in a closed, *filled*, system, the hydraulic gradient is horizontal and in an open system it lies at the surface of the stream. The total energy line has been defined as the locus of points lying one velocity head above the hydraulic grade line. The relationship of frictional losses, changes in velocity, pressures, hydraulic gradient and total energy line are shown in Fig. 47 (Skeat 1958).

The total energy, both potential and kinetic, of any given volume of water, referred to the datum plane and expressed in terms of head or elevation, is $h + Z + (V^2/2g)$ or $H + (V^2/2g)$, where h = head, Z = elevation of the pipe, v = velocity, g = acceleration of gravity, and H = level

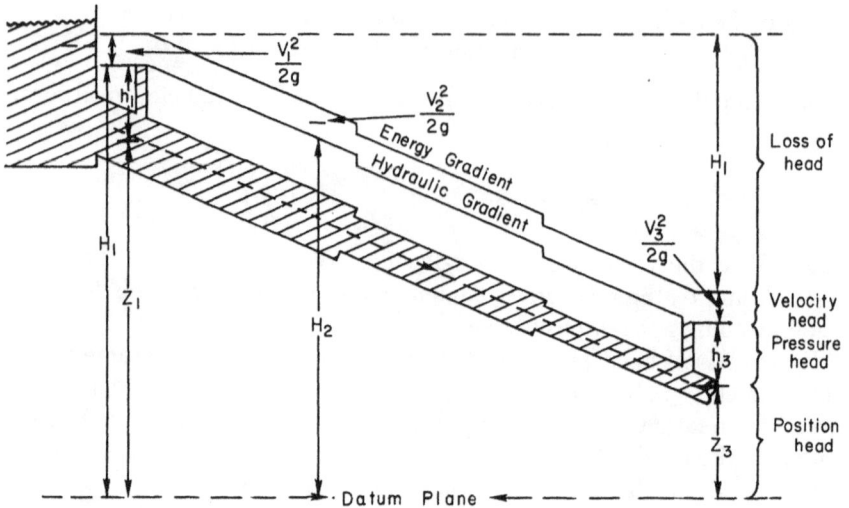

FIG. 47. BERNOULLI'S THEOREM APPLIED TO FLOW IN PIPES

of the hydraulic gradient. In the ideal case where no loss of energy occurs in going from one point to another, Bernoulli's theorem is applicable in the form $h + Z + (v^2/2g) =$ a constant. However, in the flow of water or any real liquid, energy is dissipated as heat due to the internal friction or viscosity of the substance so that the value of the equation (the remaining amount of energy) comes smaller as the water flows along the pipe. Letting h_f represent the loss of head between two points, the equations

$$h_1 + Z_1 + \frac{v_1^2}{2g} = h_2 + Z_2\frac{v_2^2}{2g} + h_f$$

can be written. If the pipe is of uniform diameter throughout, the velocities must, of necessity, be equal at all points, so that $h_1 + Z_1 = h_2 + Z_2 + h_f$. Referring to Fig. 47 with this equation in mind, it is seen that the difference in elevation of the hydraulic gradient at any two points represents the head lost in friction. Since the elevation of the grade line is independent of the pipe elevation, it is convenient to refer directly to the datum plane and write $H_1 = H_2 + h_f$. The head required ($H_1 - H_2$) to overcome friction between any two points in a pipe causing velocity equal to v, is given by:

$$H_1 - H_2 = \frac{v_1}{2g} - \frac{v_2}{2g} + h_f$$

The hydraulic grade line is employed in system design as an aid in calculating the pipe dimensions necessary to insure adequate strength and rate of flow at all points.

MATERIALS FOR PIPES

Cast iron is well-suited for conduits of moderate size and for lines where frequent use of branches and special fittings is anticipated. This material is resistant to corrosion and is fairly easy to install. Pit-cast iron pipe is available in standard lengths of 12 ft., 16 ft., and five meters. Centrifugally cast pipe, which has a slightly smoother bore than the other variety and somewhat greater strength per unit weight, can be obtained in lengths of 12, 16, 16$^1/_2$, 18, and 20 ft. Wrought-iron pipe is less durable and more expensive than that made of cast-iron, but is thinner and more easily worked. For high pressure applications, or for very large tubes, steel may be the material of choice. Cement, cement-asbestos, and bitumen-impregnated fiber are non-corroding substances which may be suitable for certain installations. Vitrified pipe can be used for small to moderate size conduits if the pressures are to be low. Copper tubing may be economical for small capacity, low pressure lines where many fittings are to be applied.

Although the material cost may be slightly higher, plastic pipe is easier to lay and has the advantage of very high resistance to corrosion. The maximum size usable without reinforcement is six inches in diameter. The principal plastics used in pipe manufacture are: acrylonitrile butadiene styrene (ABS), cellulose acetate butyrate (CAB), polyethylene, and polyvinyl chloride (PVC). Use of plastics is restricted by temperature and pressure limitations as well as by price. Some of the types have advantages of flexibility and transparency and they are not subject to the corrosion and tuberculation that occurs in metal pipes. They do not contribute harmful ions to the water and contain no crevices in which bacteria may lodge and multiply. CAB-pipes are not to be used for potable water (Poux 1962).

Reinforced plastic pipe is available in diameters from 2 to 42 in. and is said to be able to handle liquids at pressures up to 150 p.s.i. and temperatures as high as 250°F. Such pipes can be joined by butt-and-strap joints, standard flanges, sleeve-type joints, or threaded couplings. Pipes in the larger diameters must be supported by special structures or placed in a specially prepared trench which minimizes shrinking or collapse of the bed.

PUMPS

Pumps may be used in food processing plants to draw water from a natural source or from a reservoir supplied by a natural source, to reinforce a system pressure which is inadequate for some processing need, or to fill an elevated tank or reservoir used for storage or pressure-balancing purposes.

There are probably several logical schemes for classifying pumps, but one of the more meaningful methods divides them into the categories of: (1) displacement pumps, with sub-classes of (a) reciprocating and (b) rotating; (2) centrifugal; and (3) jet or ejector. The last type is not common in food processing plants. Some types of pumps, such as the bucket-and-chain, are not included in these categories.

TABLE 18

PUMP CHARACTERISTICS

Name or Class	Type of Action	Advantages and Disadvantages	Some Applications
I. Positive Displacement			
A. *Reciprocating*			
Direct-acting	A common rod connects steam piston and pump piston	Pressure fluctuates. Each stroke dispenses a given amount of fluid. Fairly efficient	Water pumping
Diaphragm	Flexing diaphragm operating in chamber with ball-type suction- and discharge-valves	Can handle viscous or gritty liquids. Controlled volume of flow. Rather inefficient	Pumps raw material slurries, sewage, corrosive solutions, etc.
B. *Rotary*			
External gear	Space between the teeth on meshing gears widens on the suction side and narrows on the discharge side	Flow not appreciably changed by variable discharge-side pressures	Tank loading and unloading
Internal-gear	A powered gear with internal teeth meshes with an idler gear having external teeth	Handles liquids of widely variable viscosity. Constant flow rates	Similar to the external gear pump
Lobe	Rotors (two or more) are lobed, not circular; may or may not be toothed	Positive pressure. Pulsating flow	Suitable for slurries, etc.
Screw	Helical rotor rolls inside a double-thread helical stator	Self-priming. Uniform flow. Can pump liquid with large suspended solids	Transferring fluids containing large particles
Vane	Relies on hinged vanes rather than gear teeth to move liquid. Single rotor	Hard to clog	Pumping suspensions, etc.
Cam-piston	Eccentrically-mounted smooth-surfaced rotor having blades in slots wiping stator surface	Generally less expensive than other rotary types. Positive action	Similar to lobe

Name or Class	Type of Action	Advantages and Disadvantages	Some Applications
II. Centrifugal			
Volute	Vaned impeller in a casing that expands progressively from inlet to outlet	Suited for low-pressure pumping. Constant flow	Transfer of abrasive fluids and most other liquids
Diffuser	Guide vaned on housing designed to reduce turbulence and increase pressure	Similar to volute but can provide greater pressure	Water pumping, etc.
Periphery	Impeller vanes generate high velocity in an annular channel	Somewhat greater pressure capabilities than diffuser	Water pumping, etc.
Axial-flow	Radially symmetrical impeller with vanes	Similar to diffuser	Water pumping, etc.
Inclined-rotor	The radii of a disc-shaped impeller are at an oblique angle to the shaft and to wall of the housing. Liquid enters axially and is discharged radially	Not quite as efficient as other varieties, but can handle problem liquids	Used for viscous slurries, pulps, gritty liquids

Displacement pumps move water by collecting it in some kind of pocket or cavity and then changing the size of the cavity so that the water is forced into the discharge outlet. The centrifugal-type of pump gives the entering fluid a high velocity through contact with a rotating vane and then changes this velocity to pressure as a result of the design of the housing and outlet. The principle of operation is entirely different from that of the rotary displacement pump and the two types should not be confused.

In reciprocating pumps, a piston or plunger moves back and forth in a cylinder, alternately drawing in fluid through the inlet valve and expelling it from the outlet valve. There are single-acting pumps in which only one end of the plunger acts on the fluid column, and double-acting pumps in which the cylinder is so constructed that the pump will act on both the forward and the return stroke. There may be 2, 3, or more cylinders. Many other variations in design are available. In all cases there is some fluctuation in pressure at the discharge valve. Priming is required at the start of operation.

Rotary displacement pumps consist of geared or lobed rotors operating in a tightly fitting housing. They give a steady pressure and do not require priming. The smooth, simple, interior construction eliminates pockets in which microorganisms can grow. Many metering pumps and sanitary pumps are of the rotary displacement design.

Centrifugal pumps are widely used for pumping water, primarily because of their low first cost, simplicity of mechanism, long life, and high efficiency. However, their efficiency is optimal only over a narrow range of head and discharge. Rate of flow cannot be regulated simply and they are of little value for metering purposes. Centrifugal pumps are not self-priming without additional mechanisms.

A more complete description of the most popular designs of displacement and centrifugal pumps is given in Table 18. The list of applications is not intended to be complete. Many types of pumps are interchangeable with little or no difference being observed in performance. Pumps for water distribution systems and those for more complex or difficult pumping functions have both been included to facilitate comparison.

There is a group of formulas called the affinity laws which expresses the theory that, for a given pump, the capacity will vary directly as the speed while the head and NPSH will vary as the cube of the speed (Horwitz 1964). The affinity laws also govern the performance of geometrically similar pumps, i.e., pumps which are identical in design except for size. If the performance of a given model pump is known, the performance of a prototype pump can be predicted. A prototype pump is made from a model by multiplying all dimensions of the model by the same factor. This procedure makes the pumps homologous, according to the terminology that is used. The size factor is denoted by the symbol K_D.

MEASURING FLOW

Parameters of value in establishing the characteristics of distribution systems are pressure, velocity, and rate of flow. Pressure is easily measured by tapping into the pipe at the appropriate point and inserting the stem of any good commercial gage. These instruments generally fall into the classes of Bourdon gages (based on the deformation of an arc-shaped or spiral tube subjected to the internal pressure of the fluid) or Schaffer gages which measure the expansion of a corrugated diaphragm. Ordinary U-tube manometers can be used for measurements over a considerable range by properly selecting the indicator fluid to be used.

Pressure readings are described as absolute, gage, or differential. Gage data (when in pounds, abbreviated as p.s.i.g.) are the difference between atmospheric pressures and the total pressures at the point of measurement. Absolute pressures include the total pressure, with no deduction being made for the atmospheric contribution. The differential pressure is the difference in pressure of a liquid at two points, usually only slightly separated, as at the two sides of a constriction in the pipe.

Rates of flow can be accurately determined by weighing the liquid discharged from an outlet over a measured period of time. However, this procedure may be very inconvenient or practically impossible. Several accurate versions of flow meters are commercially available. These have been classified as (1) inferential meters, in which the water actuates a screw, a vane, or some other inertia-dependent mechanism, and (2) the positive displacement type in which a definite volume of water is allowed to pass during each complete cycle of the mechanism. Both are accurate to a few per cent of the total reading at high rates of flow. At low rates of flow, the displacement meters are generally more reliable.

The Venturi meter can be used to measure flow rates although continuous recording of the results is inconvenient. It and the orifice meter are based on Bernoulli's theorem. The orifice meter consists of a thin flat plate with a sharp-edged circular hole that is concentric with the pipe. It is inserted in the pipe so that it is perpendicular to the stream line. A differential manometer is tapped into the pipe with one arm on each side of the orifice plate.

The Pitot tube is sometimes encountered in practical hydraulic applications. Two tubes, each with a small orifice at the end, are placed into the pipe close together but with one pointing directly upstream and the other pointing at right angles to this direction. The orifices are connected by small diameter tubes to sensitive manometers. The difference in pressures at points just inside the orifices are proportional to the velocity of the stream according to the relationship $C(V^2/2g)$. In well-designed tubes, the factor C approximates unity.

VALVES

A wide variety of valves are available to the food industry and care should be taken to select the one particular kind best adapted to the specific function which is under consideration. Sanitary valves should be capable of being rapidly disassembled and cleaned, and should have smooth internal surfaces without crevices or blind spots. For close control of flow, diaphragm, plug, or needle valves are most suitable (see Table 19).

BEGINNING SERVICE

The final design of the system, as presented by the contractor, should include a diagram showing the exact location and dimensions of the pipe, the kind of pipe depth of trenches, the radius and length of each curve, location and amount of angles and bevels, position and size of valves, and the location and description of all appurtenances such as pumps. This diagram should be superimposed upon a map or architect's drawing of the premises which will be supplied by the system.

Before a new distribution system, is put into use, it is desirable to flush and drain the lines at least twice and then to disinfect the pipes. Although

TABLE 19

VALVE CHARACTERISTICS

Name or Class	Method of Action	Advantages and Disadvantages	Typical Applications
A. *Check Valves*	Actuated by water flow		
Swing-	Disc pivots to open in direction of flow	Little pressure drop. Straight flow with little restriction at the seat	In off-on pumping systems to prevent backflow
Lift-	Opened by line pressure, closed by gravity	Tight shutoff in gas, steam, and air. Appreciable pressure drop	Gas, steam, and air lines
Ball-	Free spheroidal element	Wear equalized by ball rotation	For suspensions, viscous fluids, and in rapidly fluctuating systems
B. *Globe Valves*			
Straight-	Seats globe parallel to the flow	Good control of flow rate. Short motion required	Sanitary systems
Angle-	Globe in a right-angle fitting	Considerable turbulence and pressure drop. Easily disassembled	Used in sanitary systems where an ell is required
Y-	Globe in joint of three-legged fitting	Somewhat less pressure drop than regular globe valve	Blowing down boilers, etc.
Needle-	Tapered disc element	Requires little force for a tight closure	Steam, water, oil, and gas lines
C. *Gate Valves*	Double-disc seat and wedge-shaped closure	Relatively unobstructed flow and small pressure drop. Not for flow control. Comparatively slow in action. Minor turbulence	Positive action against high pressures
D. *Plug Valves*			
Cylinder-	Rotation of pierced plug shaped like a truncated cone	Rapid action. Can be used for semiquantitative flow control. More than one channel in plug permits directing flow into different lines	Sanitary systems
Ball-	Rotation of pierced sphere	Easily disassembled. Can be completely opened or closed with 90° turn. Uninterrupted flow when open and positive shutoff when closed	Sanitary systems
E. *Butterfly Valves*	Pivoted vane	Combines low turbulence and resistance of flow of gate valves with the positive closure of the globe valve	Flow control of slurries and other liquids

Name or Class	Method of Action	Advantages and Disadvantages	Typical Applications
F. *Diaphragm Valves*	Diaphragm flexed against seat	Often used for quantitating flow. Gives a tight closure. Turbulence and blind spots minimized. Relatively short life	Liquid or gas control

the latter step is frequently omitted in relatively small systems, particularly if chlorinated water is being received from a municipal plant, it is required when a complex network utilizing a natural water supply is put into operation. Local codes may govern the construction and hook-up of subsidiary distribution systems.

The American Water Works Association recommends the following steps be taken during the construction of municipal distribution networks.

(1) Keep the interior of the pipe clean during construction.

(2) Swab the pipe interior with an effective bactericide.

(3) Block the open ends of the pipeline to prevent water entering from the trenches.

(4) Flush the main with water having a velocity of at least two feet per second.

(5) Chlorinate with chlorine, chlorine water, or some commercial preparation which will provide the necessary concentration of chlorine.

(6) Apply chlorine or a solid disinfectant.

(7) Apply disinfectant at one end of a section with bleed-off at the other end. This treatment can be done in one of the following ways, given in order of preference.

(a) Chlorine gas-water mixture is applied at the beginning of a valved section of the line at such a rate that the chlorine dose shall not be less than 50 to 100 p.p.m. (chlorine solutions stronger than 100 p.p.m. may damage the lining if left in contact 45 min. or longer) with a slow flow of water into the pipe. The treated water is allowed to remain in the pipe for at least three hours at the end of which time the chlorine residual is not to be less than five parts per million. The pipe is thoroughly flushed and tested for adequacy of disinfection. If necessary, it is rechlorinated.

(b) A solution of calcium hypochlorite, chlorinated lime, or a proprietary equivalent may be used in place of chlorine gas. After the pipe has been subjected to a preliminary flushing, a five per cent solution of any of these substances should be introduced into the pipe in a manner similar to that used for the addition of gas.

(c) Dry calcium hypochlorite or chlorinated lime may be used in a procedure such that there is at least one pound of the disinfectant containing 70% available chlorine for every 1,680 gal. of water in the pipe. The dry material can be shaken into the suction well of a pump connected to the entrance of the pipe, or it may be distributed through each length of the pipe as it is being laid. After the hypochlorite has been introduced, water is admitted very slowly so as to avoid washing it all to the end of the line. After the chlorination is complete, in a few hours, the pipes must be flushed until the odor of chlorine disappears, or until the residual chlorine determination is within acceptable limits.

These rules for municipal systems can serve as guides or be adapted for the disinfection of smaller plant distribution networks.

In case of serious contamination of complete systems, as by flood waters, similar programs of flushing and disinfection may be necessary.

CLEANING

After prolonged usage, deposits may accumulate on the inside of pipes to an extent sufficient to reduce the flow to an unsatisfactory level. These incrustations are usually composed of calcium carbonate although corrosion products from the pipe may also be found. The time required for this situation to develop depends upon several factors but principally upon the hardness of the water (and whether the hardness is of a temporary or permanent type), the pipe material, the rate of flow, and temperature changes within the system. Although cleaning of the pipes is neither simple nor inexpensive, its value in improving capacity of the lines can often outweigh the disadvantages.

In small pipes or in isolated segments of a distribution system, acids can sometimes be used for the removal of calcium carbonate deposits. The cheapest commercial forms of sulfuric or hydrochloric acids are generally satisfactory. To reduce the destruction of pipe material by the acid, inhibitors can be added to hold the hydrogen gas which is initially evolved in a film on the surface of the metal and thereby restrict access of the acid. There are commercially available products which can be used for this purpose. Some materials which have been recommended in the past are bran, flour, glue, aniline, pyridine, and quinolidine.

If the pipe is large enough and does not include any sharp turns, it can be cleaned by pushing or pulling a scraping device through it. There are specially designed machines which fit tightly into the pipe and are driven by the water pressure behind them. Such devices travel about four miles per hour and require a pressure of about 65 p.s.i. when cleaning a 12-in. pipe (Babbitt et al. 1962). Rubber balls covered by chain mesh or the like have also been used for cleaning. A scraper can also be pulled through the line by a cable which has been pushed from the entrance to the outlet by thin jointed rods.

The carbonate deposits afford a certain amount of protection to the pipe since relatively thin layers inhibit diffusion of corrosive solutes from the fluid to the metal surface. When the deposit is removed, not only is the access of corrosive substances facilitated, but the metal may be scraped free of corroded layers leaving a fresh, reactive metal surface exposed. Deterioration of tubes can be rapid under these circumstances. The pipe can be protected by the application of phosphate glasses (highly polymerized phosphates) added in the form of a 50 to 100 p.p.m. solution.

BIBLIOGRAPHY

ADDISON, H. 1945. A Treatise on Applied Hydraulics. 3rd Edition. John Wiley and Sons, New York.

ANON. 1942. Code for Pressure Piping. American Society of Mechanical Engineers, New York.

ANON. 1948. Steel Piping Materials. American Society for Testing Materials, Philadelphia.

ANON. 1960. Tentative AWWA Standard for Fabricated Electrically Welded Steel Water Pipe. American Water Works Association, New York.

ANON. 1961A. American Standard for Vertical Turbine Pumps. American Water Works Association, New York.

ANON. 1961B. American Water Works Standard for Gate Valves for Ordinary Water Works Service. American Water Works Association, New York.

ANON. 1962. Water Treatment and System Handbook. National Association of Domestic and Farm Pump Manufacturers, Annapolis, Md.

BABBITT, H. E., DOLAND, J. J., and CLEASBY, J. L. 1962. Water Supply Engineering. Sixth Edition. McGraw-Hill Book Co., New York.

BINDER, R. C. 1943. Fluid Mechanics. Prentice-Hall, Englewood Cliffs, N. J.

CROCKER, S. 1945. Piping Handbook. McGraw-Hill Book Co., New York.

DAVIS, C. V. 1942. Handbook on Applied Hydraulics. McGraw-Hill Book Co., New York.

HIRSCH, A. 1945. Manual for Water Plant Operators. Chemical Publishing Co., New York.

HOROWITZ, R. P. 1964. Affinity laws and specific speed in centrifugal pumps. Water and Sewage Works 111, 343–344.

KING, H. W. and BRATER, E. F. 1963. Handbook of Hydraulics: For the Solution of Hydrostatic and Fluid-Flow Problems. 5th Edition. McGraw-Hill Book Co., New York.

LINSLEY, R. K., JR., and FRANZINI, J. B. 1955. Elements of Hydraulic Engineering. McGraw-Hill Book Co., New York.

MILLEVILLE, H. P., and GELBER, P. 1964. Sanitary design of food processing equipment. Food Processing 25, No. 10, 93–102.

PAYROW, H. G. 1941. Sanitary Engineering. International Textbook, Scranton, Penna.

Poux, R. N.   1962.   Plastic pipe.   J. New England Water Works Assoc. *76*, 22–29

Skeat, W. O. (Editor).   1958.   Manual of British Water Supply Practice.   W. Heffer and Sons, Cambridge, England.

Williams, G. S., and Hazen, A.   1920.   Hydraulic Tables. 3rd Edition.   John Wiley and Sons, New York.

CHAPTER 12

Water and the Growth of Food-Yielding Vegetation

INTRODUCTION

The flow of aqueous solutions through the tissues of fruit-, vegetable-, and grain-producing plants controls and mediates the reactions making up their growth and maturation phases. The amount of water available to the plant from the soil and, to a lesser extent, directly from the atmosphere, is perhaps the most important single factor affecting the rates of these reactions, although the supply of energy as sunlight, the temperature, and the availability of trace elements and nitrogen are also controlling factors in many instances. In any case, restrictions on maximum yield of food organs exist if the supply of water is sub-optimal, or if the organism cannot absorb and utilize the water at a rate adequate to match the energy uptake. Not only the yield but the composition of the useful portion of the plant as well is controlled by the availability of water at various stages of the growth process.

In this chapter, the soil-plant-water relationships which interest the food technologist because of their effect on the raw materials he encounters will be surveyed briefly. More detailed treatments can be found in the recent review by Kozlowski (1964) and in textbooks on plant physiology, such as that of Meyer and Anderson (1952).

The amount of water available to a plant at any given time depends upon a complex of soil and atmospheric conditions as well as the status of the plant itself. The availability of soil water can be regarded as a competition between the plant and the soil for the moisture in their immediate area of contact, plus the speed of transfer, as vapor or liquid, from regions of relative surplus in the soil to the deficit region at the root-soil interface.

The interstitial spaces of a soil will occupy about 30 to 50% of the total volume, being near the lower figure in sandy soils and close to the higher limit in clays. In thoroughly air-dried soils (containing from less than 1 to about 5% moisture removable by oven-drying at 221°F.), these spaces may be occupied entirely by air. In saturated soils, they will be filled entirely by water. Usually, they are occupied by both air and water in varying percentages. Except in unusually dry soils, the air in the interstitial spaces will be saturated with water vapor.

The percentage of moisture in a soil which contains the maximum amount of water in equilibrium with gravitational and capillary forces is called the field capacity. It is defined as the per cent of moisture in the soil (on a dry weight basis) 2 or 3 days after a thorough wetting of the soil profile by rain or irrigation water. The water is assumed to have reached an equilibrium between capillary and gravitational forces under these conditions. Only the smaller interstices of the soil will be filled with water at field capacity. Complete saturation of the soil through filling of all of the spaces for any appreciable length of time is quite harmful to most plants.

The field capacity of a soil composed of large particles of relatively uniform size, for example, very sandy soils, will be about 5 to 8%, while some clay loams will run as high as 35 to 40%. Measurements of field capacity can be made directly or it can be approximated by measuring the water retained when a saturated sample of soil is centrifuged at 1,000 gravities—a purely empirical relationship. In all cases cited in this chapter, water content is expressed as a percentage of the dry weight of the soil.

WATER ABSORPTION AND TRANSPORT

Although liquid water may be absorbed if it comes into contact with the aerial parts of a plant, by far the greatest proportion of absorption occurs in the roots and especially through the epidermal cells and root hairs located near their terminal sections. The number of tips in a root system is said to be one of the most important indices of the latter's effectiveness as an absorbing organ. The water absorption of plants is a passive process, the demand being determined by internal water deficits resulting primarily from transpiration. It may vary greatly in response to changes in temperature or it may change very little, depending on the species.

Generally speaking, capillary movement of water in soils at or below field capacity is negligible. As a consequence, the area around the absorbing root surfaces quickly becomes reduced in water content. However, the root tips are more or less constantly growing and this brings the root surface into contact with fresh, high-moisture areas of the soil. Furthermore, absorption is rapid when the soil receives additional water from irrigation, rain, or other precipitation and continues rapidly until the condition of field capacity is reached. The roots either stop growing or reduce their rate of extension greatly when the soil is soaked. This cessation of root growth is due primarily to displacement of the air from the interstices of the soil and interference with the respiration of the roots. Poor soil aeration, high moisture tension, or high osmotic pressure of the aqueous phase of the soil (salinity) may reduce the capacity of roots to absorb so that moisture tension develops in the plant as the result of even minimal transpiration. If this impairment is sufficiently great, the plants may lose their turgidity even though the supply of liquid water seems adequate.

Water absorbed by the roots passes through several peripheral tissues before it reaches the xylem, through which transfer to other organs of the plant takes place. The xylem is a complex tissue consisting of anastomosed tracheids or tracheae (types of vessels extending continuously throughout all organs of the plant) and usually also of woody fibers and parenchyma cells. Anatomically, in woody stems it occupies a relatively thick layer between the innermost core of pith and the more external tissues such as cambium, phloem, etc. The transport vessels in herbaceous plants may be interspersed between other types of tissues. In these tubular spaces formed from degenerated cells, water moves as a continuous liquid phase.

The exact mechanism supporting transport of water against the force of gravity, as to the tops of tall trees, has been a matter of conjecture for many years. According to Kozlowski (1964), the cohesion theory is today the most widely accepted explanation of sap ascension. Very briefly, the cohesion theory states that loss of water from the leaves as a result of transpiration sets in motion the upward movement of sap and that ascending columns of water exist because of the continuity and cohesion of the substance in tubes of small diameter which are protected from the entry of air. It should not be thought that the cohesion theory is universally accepted, however.

Most authorities agree that water uptake by plant tissues is passive, water molecules merely following the osmotic pressure gradients between the cellular and extra-cellular fluids without the intervention of a specific transfer reaction. Of course, the osmotic gradients exist partially as a result of active transport of ions and other osmotically active substances through semi-permeable membranes as well as the formation and breakdown of metabolites.

Some evidence for an active mechanism of water uptake has been adduced by other investigators. For example, it appears that increases in the water content of plant tissues are dependent upon aerobic conditions, are checked by respiration inhibitors, and are promoted by auxins. Those who advocate an entirely passive role for water regard these relationships as indirect, the immediate effect of these agencies being a change in the solute concentration within the cell as a result of variations in its metabolic activity.

The factors affecting water uptake by the roots and transport of the resultant aqueous solutions through the plant under any given set of conditions depend upon inherent plant characteristics as well as the water activity in the soil. The area of root surface capable of water absorption, the efficiency of water absorption per unit of root surface, and the rate at which the aqueous solution is transported away from the root system to other tissues are factors entering into the capacity for utilization of water present in the soil. In turn, these are determined by the genetic make-up of the plant, its previous history, and environmental conditions existing at the time.

The environmental conditions other than soil conditions affecting water uptake are the temperature, the intensity and spectral distribution of sunlight reaching the plant, the relative humidity, and air velocity. These conditions govern the rate at which water is lost from the plant through transpiration. Water is also consumed by development of new cells or by tissue enlargement but to an extent which is small in comparison to that lost through transpiration. The rate of consumption of water is the principal governing factor in determining the rate of water removal from the root system of a healthy plant when a surplus of soil water is available.

As might be expected, low soil temperatures impede absorption of water by the roots. The effect is most noticeable in soils having a moisture content below field capacity. Decreasing the temperature increases the viscosity of aqueous solutions and decreases cell permeability, both factors operating to slow the rates at which water moves to the root surface and is taken up by the root cells.

Living plants lose water to the atmosphere in a process called transpiration. By far the greatest percentage of transpiration occurs from the leaves, but any of the aerial parts of the plant may participate in this process, and roots, also, if they are exposed to the air. Most leaf transpiration takes place through stomates, small surface pores of fluctuating diameter which communicate with spaces in the interior of the leaf. The cells lining these spaces lose water by evaporation and the water vapor diffuses through the stomates.

The rate of transpiration is controlled by the status of the plant and by such environmental factors as the intensity of solar radiation, humidity, temperature, wind, soil conditions influencing the availability of water, and atmospheric pressure. Common rates of transpiration in plants growing in temperate regions will be about 50 to 250 gm./sq. m./hr. When conditions unfavorable to transpiration occur, and especially at night, transpiration of a normal healthy plant may fall to 10 gm./sq. m./hr. or even less. The transpiration potential in most plants exceeds the absorption potential and, after continuous and prolonged transpiration, a water deficit will exist in the tissues. An herbaceous plant, for example, can lose several times its own volume of water in a single day, under favorable conditions. At these high rates of transpiration, transient wilting will occur even in soils of high moisture content. Ordinarily, the cells in these plants will recover their turgor during the night. The water content of an entire plant is principally a function of the relative rates of absorption and transpiration of water although these may be limited by the genetic constitution of the plant cells.

Peters (1960) showed that plants held in atmospheres of low relative humidity had their growth strongly influenced by soil-moisture tension and moisture content of the soil. When the plants were placed in atmospheres of high water activity, growth was relatively insensitive to the water activity of the soil. He interpreted the findings as indicating that the total free energy path causing water transport should be considered in analyzing effects of water deficit on growth.

Below some moisture content, which may be different for each soil and plant system, the rate at which soil water can be taken in by the roots becomes inadequate to replace that lost by transpiration. Under these conditions, most plants will wilt and, if the water supply does not improve or the rate of water loss decrease, they will die. This soil moisture content is called the permanent wilting percentage and, for most plants, lies between about 3 and 23%, depending on the type of soil. The lower limit of this range includes coarse sandy soil and the higher limit is applicable to some fine-grained clays, while silty loams are intermediate. According to Wadleigh (1955), the moisture held by a sample of soil against a force of 15 atm. (221 p.s.i.) is a good approximation of the permanent wilting percentage. It can be determined by growing dwarf sunflower plants in small pots of the soil held at decreasing moisture percentages.

Restrictions on growth begin to occur at soil-moisture contents considerably above the permanent wilting percentage. The overall effect of internal water stress is growth inhibition.

It appears that plants can absorb some water from the soil at or below the permanent wilting percentage although vegetative growth is probably inhibited completely. Staple and Lehane (1941) found that uptake of water by wheat plants under conditions of permanent wilting affected the yield and quality of grain.

The sequence of events leading to the formation of a usable food organ by a plant can be divided into growth, maturation, and ripening phases. The growth phase, during which the food develops to substantially its full size, is the result of two processes: cell division (production of new cells), and cell enlargement. There is ordinarily little or no increase in the amount of protoplasm as plant cells enlarge. Instead, the cytoplasm of the original small cell is gradually spread over a larger area of cell wall. The principal factor involved is a considerable increase in size of the vacuoles due to an influx of water. Cell walls may increase in thickness, requiring additional water for hydrating the greater amount of structural material.

Soil moisture deficits inhibit cell enlargement but have little effect on cell division until the permanent wilting percentage is reached. Through osmotic and other effects, a living plant cell tends to absorb water from its surroundings until a positive pressure on the cell wall is established. This turgor pressure is the main force leading to cell enlargement, according to several authorities, and plant growth is directly related to the degree and duration of turgescence of the tissues (Wadleigh 1955). Its effectiveness is controlled by the action of the plant hormone auxin. Other authorities regard turgor pressure as a secondary effect not directly related to cell enlargement.

It appears that some or most plants respond to increases of osmotic pressure in the medium by increases in osmotic pressure of the contained fluid. Bernstein (1962) showed that essentially complete osmotic adjustment of all plant parts to increased osmotic pressure of the medium occurred in time. Leaves, stems, and roots increased by about the same amount as the osmotic pressure of the medium was increased. In bean and pepper plants, osmotic pressure increased at maximum rates of about 0.5 to 1.0 atm./day when they were exposed to 1.5 atm. additions of sodium chloride.

There is a reduction in the amount of water taken up as the osmotic pressure increases. Transpiration of tomato plants was reduced about half by changing them from a solution of 0.5 to 2 atm. (Army and Kozlowski 1951). Uniform corn plants placed in substrates varying in osmotic pressure from 0.8 to 4.8 atm. showed a marked decrease in absorption with increasing osmotic pressures (Hayward and Spurr 1944). Absorption at a given osmotic pressure was about the same whether the solute was organic (sucrose or mannitol) or inorganic (sodium chloride, sodium sulfate, or calcium chloride). Initially, decrease of water uptake is due to a lessening of the diffusion pressure gradient from the soil to the roots. As adaptation to the higher soil concentration of solutes in the soil occurs, there is reduced root and shoot growth, decreased root permeability, and fewer root hairs (Hayward and Blair 1942).

There are two methods which have been rather widely used for estimating the osmotic pressures of plant cells and tissues: (1) the plasmolytic method and (2) the cryoscopic method. In the cryoscopic method, a sample of fluid is pressed from a plant tissue and its freezing point is determined by the usual procedures. The freezing point depression is related to the osmotic pressure in a known manner, enabling the calculation of the latter from these data. The osmotic pressure in atmospheres $= 12.04$ times the freezing point depression in degrees Centigrade.

The plasmolytic method requires the immersion of replicate strips or sections of tissue in a series of solutions of an osmotically active substance such as sucrose. The concentrations of the solutions are varied in a series of equal steps (e.g., 0.02 to 0.05 moles per step). The total range of concentrations is selected to exceed the limits of the range in which the average osmotic pressure of the cell sap is expected to lie. The strips of tissue are observed microscopically after remaining in the solutions for 30 minutes or longer. The cells in the most concentrated solutions will become severely plasmolyzed, the cytoplasm pulling away from the cell wall and exhibiting other typical changes. In the most dilute solutions, no plasmolysis will be seen. At some intermediate concentration, about half the cells will be plasmolyzed while the remainder will appear unaffected. The osmotic pressure of this solution can be taken as the average osmotic pressure of the cells, although, as several authors have remarked, it would be more accurate to call it the osmotic pressure at incipient plasmolysis.

The difference between the osmotic pressure at incipient plasmolysis and the true osmotic pressure of the cells may be as much as 3 or 4 atm. (Meyer and Anderson 1952). This difference is due to the shrinkage of cells which usually occurs before plasmolysis is observed. If the volumes of the cells can be estimated reasonably accurately by using microscopic measurements, the contribution of shrinkage to the data obtained by the plasmolytic method can be corrected. The following equation is used:

$$M_0 = \frac{M_1 V_1}{V_0}$$

in which M_1 is the concentration of sucrose at incipient plasmolysis as determined in the plasmolytic method, V_1 is the average volume reached by cells in this solution, M_0 is the sucrose concentration at the original cell volume, and V_0 is the original volume of the cell.

The plasmolytic method has been widely adopted in spite of some rather formidable technical difficulties and theoretical objections. One of the major problems of the technique is identification of plasmolysis in unpigmented cells.

The osmotic pressures of different plants, or of different tissues from the same plant, may vary widely. Most plant cell values will be in a range of about 4 to 30 atm.

The internal competition for water may result in translocation in a reverse direction when the total supply becomes inadequate. For example, plant parts may transfer part of their moisture content to the leaves during periods in which transpiration materially exceeds absorption. The internal pressure gradient set up in this way causes a decrease of turgor pressure and a shrinkage of plant parts as water is withdrawn from them. Even relatively rigid parts such as tree stems are affected. Apples, cherries, plums, and walnuts have been shown to lose moisture to leaves, with resultant shrinkage. Drought conditions, seasonal fluctuations in moisture availability, and the daily change in transpiration rates will affect size. Furr and Taylor (1933) described cases of decreased fruit growth resulting from internal moisture stress while other workers found that leaf suction could cause oranges to decrease 25 to 35% in moisture content during hot and dry weather. Several other authors have reported that citrus fruits, avocados, pears, and other fruits decrease in diameter during the daytime but recover turgidity and increase in diameter at night. It would appear that the yield of such fruits would be best if they were picked in the early morning hours. Field drying of legumes, cereals, and the like is, in part, caused by such reverse translocations of aqueous substances.

Bartholomew (1926) studied the diurnal fluctuations in diameters of lemons attached to the tree. The lemons started to shrink about 6:00 A.M. each day, the trend being reversed at about 4:00 P.M. The period of shrinkage corresponded to the time when transpiration was causing heavy losses of water from the leaves. Water moved from the fruits to the leaves under these conditions. The osmotic pressures of green fruits are almost always less than those of leaves on the same plant. This relationship may be reversed as the fruit ripens, but such a change is infrequent.

The cutting of narrow rings around the branches of apple trees has been practiced by some growers in attempts to increase the size of the developing fruits. This operation interrupts the tissue pathways through which soluble carbohydrates return to the lower parts of the tree. Water and substances synthesized by leaves distal to the ring will remain in the ringed branch and may be taken up by the developing fruits on it. The greater supply of substrate appears to favor increases in the size of the fruit. No permanent harm is done to the plant if the gap in the phloem is narrow enough so that it can be bridged by wound tissue.

BIBLIOGRAPHY

ANON. 1954. Diagnosis and improvement of saline and alkali soils. U. S. Dept. Agr. Handbook 60.

ARMY, T. J., and KOZLOWSKI, T. T. 1951. Availability of soil moisture for active absorption in drying soil. Plant Physiol. 26, 353–362.

BARTHOLOMEW, E. T. 1926. Internal decline of lemons. III. Water deficit in lemon fruits caused by excessive leaf evaporation. Am. J. Bot. 13, 102–117.

BERNSTEIN, L. 1962. Salt tolerance of plants and the potential use of saline waters for irrigations. Nat. Research Council Publ. 942, 273–283.

BERNSTEIN, L. 1963. Osmotic adjustment of plants to saline media. II. Dynamic phase. Am. J. Botany 50, 360–370.

BERNSTEIN, L., and HAYWARD, H. E. 1958. Physiology of salt tolerance. Ann. Rev. Plant Physiol. 9, 25–46.

BULL, H. B. 1943. Physical Biochemistry. John Wiley and Sons, New York.

CRAFTS, A. S., CURRIER, H. B., and STOCKING, C. R. 1949. Water in the Physiology of Plants. Chronica Botanica, Waltham, Mass.

EZELL, B. D., WILCOX, M. S., and DEMAREE, K. D. 1956. Physiological and biochemical effects of storage humidity on sweet potatoes. J. Agr. Food Chem. 4, 640–644.

FLANNERY, W. L. 1956. Current status of knowledge of halophilic bacteria. Bacteriol. Rev. 20, 49–66.

FURR, J. R., and TAYLOR, C. A. 1933. Osmotic relationships of plants. Proc. Am. Soc. Hort. Sci. 30, 45–50.

GARDNER, W. R., and NIEMAN, R. H. 1964. Lower limit of water availability to plants. Science 143, 1460–1462.

HAINES, W. B. 1927. Studies in the physical properties of soils. IV. A further contribution to the capillary phenomena in soils. J. Agr. Sci. 17, 264–290.

HAINES, W. B. 1930. Studies in the physical properties of soils. V. The hysteresis effect in capillary properties and the modes of moisture distribution associated therewith. J. Agr. Sci. 20, 97–116.

HAYWARD, H. E., and BERNSTEIN, L. 1958. Plant growth relationships on salt-affected soils. Botan. Rev. 24, 584–635.

HAYWARD, H. E., and BLAIR, W. M. 1942. Responses of Valencia orange seedlings to varying concentrations of chloride and hydrogen ions. Am. J. Bot. 29, 148–155.

HAYWARD, H. E., and SPURR, W. B. 1944. Effect of isosmotic concentrations of inorganic and organic substrates on entry of water into corn roots. Bot. Gaz. 106, 131–139.

INGRAM, M. 1957. Microorganisms resisting high concentrations of sugars or salts. Symposium Soc. Gen. Microbiol. No. 7, (Microbial Ecology) 90–113.

JACKSON, R. D., ROSE, D. A., and PENMAN, H. L. 1965. Circulation of water in a soil under a temperature gradient. Nature 205, 314–316.

KEMPER, W. D., ROBINSON, C. W., and GOLUS, H. M. 1961. Growth rates of barley and corn as affected by changes in soil moisture stress. Soil Sci. 91, 332–338.

KOZLOWSKI, T. T. 1964. Water Metabolism in Plants. Harper and Row, New York.

KRAMER, P. J. 1951. Causes of injury to plants resulting from flooding of the soil. Plant Physiol. 26, 722–736.

KRAMER, P. J. 1959. Transpiration and the water economy of plants. In Plant Physiology, Volume II. Edited by F. C. Steward. Academic Press, New York.

KRAMER, P. J., and KOZLOWSKI, T. T. 1960. Physiology of Trees. McGraw-Hill Book Co., New York.

LUNDEGARDH, H. 1950. Translocation of salt and water through wheat roots. Physiol. Plantarum 3, 103–151.

MASTOVSKY, J., and KAREL, V. 1961. A study of the possibility of shortening dormancy of barley and reducing its water sensitivity. Am. Brewer 94, 41–44.

McILRATH, W. J., and DRUMKE J. S. 1961. Investigation of plant water relations of selected food crops as affected by various physiological conditions. Rept. No. 4, Project No. 7-84-06-031a. Quartermaster Food and Container Institute for the Armed Forces.

MEYER, B. S., and ANDERSON, D. P. 1952. Plant Physiology, Second Edition. D. Van Nostrand Co., Princeton.

PETERS, D. B.   1960.   Growth and water absorption by corn as influenced by soil moisture tension, moisture content, and relative humidity.   Soil Sci. Soc. Am. Proc. *24*, 523–526.

PHILLIPS, I. D. J.   1964.   Root-shoot hormone relations.   II.   Changes in endogenous auxin produced by flooding of the root system in *Helianthus annuus*. Annals Bot. *28*, 38–45.

SAUNDERS, J. F.   1965.   The water molecule in biological systems.   Science *147*, 179–181.

SCOTT, G. E., HEYNE, E. G., and FINNEY, K. F.   1957.   Development of the hard red winter wheat kernel in relation to yield, test weight, kernel weight, moisture content, and milling and baking quality.   Agron. J. *49*, 509–513.

SCOTT, W. J.   1962.   Available water and microbial growth.   *In* Proc. Low Temperature Microbiology Symposium (1961).   Campbell Soup Co., Camden, N. J.

SHELLENBERGER, J. A.   1959.   Wheat.   *In* The Chemistry and Technology of Cereals as Food and Feed.   Avi Publ. Co., Westport, Conn.

SLATYER, R. O.   1961.   Effects of several osmotic substrates on the water relationships of tomato.   Austral. J. Biol. Sci. *14*, 519–540.

SLAYMAN, C. L., and SLAYMAN, C. W.   1962.   Measurement of membrane potentials in *Neurospora*.   Science *136*, 876–877.

STAPLE, W. J., and LEHANE, J.   1941.   The use of the wilting coefficient in soil moisture studies in southwestern Saskatchewan.   J. Sci. Agr. *21*, 440–447.

STEWARD, F. C.   1959.   Plant Physiology: A Treatise.   Vol. II.   Plants in Relation to Water and Solutes.   Academic Press, New York.

STONE, J. F., SHAW, R. H., and KIRKHAM, D.   1960.   Statistical parameters and reproducibility of the neutron method for measuring soil moisture.   Proc. Soil Sci. Soc. Am. *24*, 435–438.

TOOLE, E. H., and HENDRICKS, S. B.   1956.   Physiology of seed germination. Ann. Rev. Plant Physiol. *7*, 299–324.

VAADIA, Y., RANEY, F. C., and HAGAN, R. M.   1961.   Plant water deficits and physiological processes.   Ann. Rev. Plant Physiol. *12*, 265–292.

WADLEIGH, C. H.   1955.   Soil moisture in relation to plant growth.   U. S. Dept. Agr. Yearbook *1955*, 358–361.

CHAPTER 13

Effect of Water Impurities on Food Characteristics

INTRODUCTION

The minor constituents of a water supply can have an appreciable effect on the appearance, texture, or flavor of foods in which it has been used as an ingredient. Even ice-making, which would seem to be about the simplest processing method applied to water, has special raw material requirements that must be observed if maximum quality is to be attained. When large blocks of ice are frozen, the minerals dissolved in the water tend to concentrate in a core of eutectic composition, that, when solidified, becomes undesirably cloudy in appearance. Removal of the core and replacement by fresh water can be resorted to as a means of obtaining uniformly clear blocks. De-ionization of the water supply is used by some manufacturers in order to reduce the number of times the core liquid must be replaced. Schemes for rating water supplies in terms of the probable quality of the ice and the treatment which would be required during freezing have been devised. These are based on the weighting (i.e., the multiplication by equivalence factors) of the salt concentration determined by analysis. Chlorides are estimated to be 75% as effective as the sulfates while carbonates are 125% as effective, for example. The weighted concentrations in milligram per liter are summed to give a number related to the treatment which will be required. The limit above which satisfactory ice cannot be made has been given variously as from 684 to 710 mg./l.

There are many products composed principally of ingredient water. Carbonated beverages, beer and ale, reconstituted dehydrated foods, soups, and many other types of foods contain more ingredient water than all other components combined. It is evident that the characteristics of these foods will be greatly influenced by the quality of the water supply.

Technologists in the beverage industry have long recognized the influence on quality of the minor components of water. The requirements for carbonated beverages are usually more exacting than those for drinking water. Many bottlers use a sand filter following coagulation in addition to a filter bed of activated carbon to further purify water from municipal suppliers. Low alkalinity is desirable (limits from 50 to 100 p.p.m. are frequently specified) in order to reduce the neutralization of the acidifying ingredient.   Reduction of alkalinity may be accomplished by several methods, but the cold lime water softening process is probably the most common.

Complete absence of microorganisms, organic matter, turbidity, color, and foreign flavors is considered desirable though seldom attained in practice.   Iron and manganese must be kept to a very low level because they tend to precipitate some of the coloring materials added to the beverage.   Ranges for threshold values in p.p.m. of some of the common quality criteria were collected and published by McKee and Wolf (1963) and are reproduced below:

Alkalinity, as $CaCO_3$............................50–128 p.p.m.
Hardness, as $CaCO_3$............................200–250 p.p.m.
Turbidity ....1.0–2.0 units
Color ....5.0–10.0 units
Taste and odor ....None-Low
Total dissolved solids............................850 p.p.m.
Iron ....0.1–0.2 p.p.m.
Manganese ....0.2 p.p.m.
Iron + Manganese................................0.1–0.5 p.p.m.
Chlorides ....250 p.p.m.
Sulfates ....250 p.p.m.
Fluoride ....0.2–1.0 p.p.m.
Hydrogen sulfide ....0–0.2 p.p.m.
Organic matter...................................Infinitesimal
Algae and protozoa...............................None
Oxygen consumed.................................1.5

BEER AND ALE

The characteristics of water available to breweries are traditionally regarded as pre-eminent factors affecting the quality of their beer.   In former times, the success of breweries in certain locations was attributed largely to the water supply.   This tradition is the basis for the emphasis on springs, lakes, and the like in many beer advertisements, allusions which are completely lost on the average consumer.   At the present time, the majority of brewers in the United States use water from the municipal system with little or no additional treatment.   It is probable that the quality of beer could be improved in many cases by suitable modification of the water supply.   Table 20 lists the composition of water available to four famous breweries.   A wide variation in several features is apparent.

The salts contributed by the water used in its manufacture are but a small proportion of the total mineral content of beer, so that their direct effect on the final flavor is slight.   However, the effect of particular ions on protein solubility and yeast metabolism, as well as other indirect actions, can be very important.

TABLE 20

COMPOSITION OF VARIOUS BREWING WATERS (MILLIGRAM/LITER)[1]

	Pilsen	Dortmund	Munich	Burton
Total solids after evaporation	51	1,110	284	1,790
Calcium as CaO	9.8	367	106	520
Magnesium as MgO	1.2	38	30	145
Sulfates as SO₄	4.3	240	7.5	756
Chloride	5.0	107	2.0	34
Nitrates as N₂O₅	Trace	Trace	Trace	22

[1] From DeClerck (1957).

Water low in alkalinity and rather high in calcium sulfate hardness is needed so that the mash pH will be low enough (4.6 to 5.7) for optimum enzyme activity.   If the supply is too soft, addition of calcium sulfate (gypsum) and sodium chloride may be required. The addition of hardening salts is called "burtonizing."   According to Ohlmeyer (1959), a high wort pH leads to uneven saccharification during mashing, difficulties in the separation of the wort from the spent grain (resulting in a low yield of extract), incomplete coagulation of the proteins during boiling, sharper bittering flavor from the hops, and a biologically unstable beer susceptible to infection with lactic organisms.

The acidifying effect of calcium from the water supply (or added gypsum) is due to the reaction of calcium ions with secondary phosphate ions derived from the malt, as indicated by the equation:

$$3 \text{ Ca}^{++} + 2 \text{ HPO}_4^{--} \rightarrow \text{Ca}_3(\text{PO}_4)_2 + 2 \text{ H}^+$$

The triorthophosphate salt of calcium has a very low solubility, but the reaction does not go to completion.

Nitrites or nitrates, the latter being partially reduced to nitrite during fermentation, may interfere with yeast activity even if present at very low concentrations.   Iron contributes an off-flavor and may discolor the beer by interfering with the clarification process.   More than one part per million of iron also has a deleterious effect on enzymes in the mashing step since it can combine with these proteins as well as other barley components.

Bonnet (1955) states that magnesium concentrations in excess of 30 mg./l. impart a reddish tint and an acridness to beer.   Chlorides are said to mellow and sweeten or soften beer, especially at levels of 20 to 60 mg./l. Higher concentrations are to be avoided.   Copper in trace amounts has been blamed for accentuating the "gushing" phenomenon (Hoak 1956). Fluoride seems to stimulate yeast metabolism at concentrations of 1 to 5 mg./l. (Webber and Taylor 1954), while inhibition has been observed at levels higher than 10 mg./l. (Slater 1951).   Silica may increase the turbidity if more than 50 p.p.m. are present, although this effect is relatively unimportant (Pozen 1940).

Bonnet (1955) and others have compiled lists of suggested limits for water to be used in brewing and some of these are given below:

Color ....0–10 p.p.m.
Turbidity ....0–10 p.p.m.
Taste and odor ....None-low
Total dissolved solids (300 for any one substance) ..500–1500 p.p.m.
pH 6.5–7.0
Alkalinity, as CaCO$_3$ ....75–80 (light beer)
                                               80–150 (dark beer)
Iron ....0.1–1.0 p.p.m
Manganese ....0.1 p.p.m.
Iron + Manganese ....0.1 p.p.m.
Carbonate ....50–68 p.p.m.
Nitrite 0
Nitrate ....10 p.p.m.
Chloride ....60–100 p.p.m.
Silica ....50 p.p.m.
Hydrogen sulfide ....0.2 p.p.m.
Fluoride ....1.0 p.p.m.
Calcium ....100–200 p.p.m. (light beer)
                                 200–500 p.p.m. (dark beer)
Calcium sulfate ....100–500 p.p.m.
Magnesium ....30 p.p.m.

DAIRY FOODS

In dairy operations, water is used mostly for cleaning, and sometimes for washing products, dilution, etc. Bacteriologically, the water should be at least of potable quality. It should be free of copper, and low in iron and manganese since these ions can act as catalysts for lipid oxidation and may lead to the development of fatty, oily, or fishy tastes in the finished products. Bonnet (1955) published the following limits for water used in dairy processing:

Dry residue ....<500 mg./l.
Nitrate, as N$_2$O$_5$ ....<30 mg./l.
Nitrite .... None
Ammonia, as NH$_3$ ....Traces only
Chlorides ....<30 mg./l.
Sulfates ....<60 mg./l.
Organic matter, in terms of KMnO$_4$ ....<12 mg.l.
Hardness, as CaCO$_3$ ....<180 mg./l.
Iron ....0.1 to 0.3 mg./l.
Manganese ....0.3 to 0.1 mg./l.
Color ....None
Odor ....None

When nonfat dry milk is used as the raw material for cottage cheese and similar products, the nature of the water added to reconstitute the dried substance affects the quality of the finished product.   Moore and Kapsalis (1963) found that prolonged rennet coagulation time and low curd tension, both undesirable in cheese manufacture, may be caused by the carbonate and bicarbonate alkalinity of the reconstitution water.   Acid development by a one per cent inoculum of lactic culture was materially inhibited by bicarbonate while a five per cent inoculum developed acid at a nearly normal rate.   Other natural water constituents, except sodium ion, seem to be relatively unimportant.   Sodium displaces calcium from the casein molecules, resulting in a higher affinity for water in the latter.   As a result, the curd forms more slowly and is softer.

CONFECTIONERY

Water used in making confectionery should be of drinking water quality as a minimum.   Low solids concentration is necessary for water used in hard candy since the very low final moisture content of these products leads to precipitation of salts with resulting cloudy appearance and fragile texture (Ingleton 1962).   It has been reported that water having a pH below 7.0, including distilled water, tends to favor inversion of the sucrose to produce a sticky piece.   Natural water with less than 50 or 100 (recommendations vary) mg./l. of the common dissolved mineral constituents is generally considered satisfactory for most varieties of confectionery. Jordan (1930) pointed out that differing compositions of water are responsible for some of the failures encountered in trying to duplicate formula results in different locations.   He suggested that use of distilled water containing a trace of added alkali would eliminate many of these unexplained fluctuations.

Upper limits which have been suggested by various authors are:

Taste and odor ....Low
Iron ....0.2 p.p.m.
Manganese ....0.2 p.p.m.
Iron + Manganese ....0.2 p.p.m.
Hydrogen sulfide ....0.2 p.p.m.
pH ....>7.0

PROCESSED FRUITS AND VEGETABLES

In the processing of fruits and vegetables, the carryover of microorganisms and of unpleasant tastes, off-odors, turbidity, and color from ingredient water are obvious factors in decreasing the quality of the finished product.   Less obvious are the changes brought about by the interaction of compounds in the food with some of the usually innocuous ions present in potable water supplies.   Probably the best understood of these reactions are the structural or textural changes which occur when calcium or certain other cations are present in the water in sufficient concentration to cause hardening in cooked foods. This firming is desirable in products such as potatoes and tomatoes. In fact, traces of calcium salts may be added to canned tomatoes to prevent structural breakdown during heat-processing. In other cases, the textural changes are disadvantageous. The phenomenon of hardening in cooked dry beans and other legumes (such as peas and lentils) has been extensively studied.

Huenink and Bartow (1915) were among the first to show that calcium ions affected the texture of cooked dry beans. When beans were canned in local water (ion concentrations in parts per million were: K 2.6, Na 29.0, NH_4 2.3, Mg 34.9, Ca 70.1, Fe 1.0, Al 1.3, NO_3 0.7, Cl 3.5, SO_4 2.3, and SiO_2 18.9) they were very much harder than those canned in distilled water. When calcium chloride was added to the distilled water in increments of 100 p.p.m. up to 1,000 p.p.m., increasing hardness was observed. Beans canned in water containing more than 300 p.p.m. calcium chloride had unacceptable texture. Results when magnesium chloride was added were similar but not as pronounced. Sodium carbonate had an opposite effect, and beans canned in water to which 1,000 p.p.m. of this compound had been added were excessively soft and had the appearance of being greatly overcooked. These authors also observed a darkening of the product related to the percentage of calcium ion.

Masters (1918) and Masters and Garbutt (1920) found that the cooking time for butter beans was less in distilled water than in London tap water of 120 p.p.m. temporary hardness and 50 p.p.m. permanent hardness. This effect has been reported for several legumes and it may reasonably be extrapolated to include all of them. Lowe (1955) described the experiences of many of her home economics classes in cooking navy beans. Cooking time was decreased about 20% by cooking or soaking in distilled water as opposed to cooking or soaking in tap water. Addition of soda in small amounts reduced the cooking time slightly and appeared to lighten the color of the beans.

The adverse texture changes observed in legumes as a result of cooking them in water containing calcium is thought to be due to an interaction of the cations and the pectic substances in the cell walls of the food. It is possible that proteins are also involved but little or no published data are available on this point. The effect of soda is probably due principally to the shift in hydrogen ion concentration, although competition of sodium ions for the calcium-binding sites cannot be ruled out.

Observations such as the foregoing have led many authorities to recommend the use of water of zero hardness for canning peas, beans, and lentils. Zeolite water softeners are widely employed in canneries which pack legumes.

SUGAR, STARCH, AND CORN SYRUP

Sugar refineries require process water of high purity. It should be as free as possible from dissolved solids and completely devoid of suspended material. Electrolytes decrease the yield of sugar by causing inversion—glucose and fructose are not separated by the usual processing steps—as do many microorganisms. Hard water may form precipitates which accumulate in the product. Maxima for some of the ions have been published by McKee and Wolf (1963):

	p.p.m.
Calcium	20
Magnesium	10
Sulfate	20
Chloride	20
Bicarbonates as $CaCO_3$	100
Iron	0.1

High quality corn starch should have a very low ash content. Since salts from hard process water tend to accumulate in this product, use of soft or de-ionized water is indicated, where possible. Iron and manganese give a yellowish discoloration to starch, and Bonnet (1955) has recommended an upper limit of one part per million for these elements in corn processing water. Some authorities have claimed that high magnesium concentrations are especially troublesome and lead to the development of cloudiness in corn syrup.

BAKERY PRODUCTS

The dissolved mineral substances in absorption (ingredient) water constitute but a very small fraction of the soluble material in bakery products. Except for certain dietary products, all bakery foods contain at least one per cent salt, and usually considerably more. This provides sodium and chloride concentrations which exceed those coming in with the water. The flour contains a fraction of one per cent of mineral substances and various ionic constituents result from the addition of milk, yeast food, dough improvers, chemical leavening agents, and other ingredients. In spite of these "competitors," the salts present in water can have an appreciable effect on the machineability of the dough as well as the appearance and texture of the finished products. Soft waters may result in sticky doughs which require less than the normal amount of absorption, but this may be overcome by increases in the use of the proper type of dough improvers or more salt.

Brooke (1939) reported that copper salts, iron salts, aluminum salts, tannins, silicates, and phosphates, in the forms and quantities found in natural water supplies, do not ordinarily affect fermentation. Evidence was found for an effect on fermentation of the following salts: calcium oxide, calcium carbonate, calcium sulfate, magnesium chloride, magnesium oxide, and sodium bicarbonate.

A systematic study of the effect on fermentation and dough quality of calcium carbonate, calcium sulfate, calcium oxide, magnesium chloride, and magnesium oxide was made by Brown (1939). Addition of magnesium and calcium salts led to increased dough stiffness. The phenomenon, which has been observed by many others, has been attributed to the reaction of gluten proteins with the cations in question, especially in cross-linking types of interactions. Brown found that sodium bicarbonate did not affect dough fermentation or loaf volumes when added at levels up to 1,000 p.p.m. Magnesium oxide was also ineffective up to 500 p.p.m. but led to decreased loaf volumes when larger amounts were added. Beneficial effects on fermentation and loaf volume were observed when 50 to 500 p.p.m. of magnesium chloride, 100 to 1,000 p.p.m. of calcium carbonate, and 50 to 300 p.p.m. calcium sulfate were used. Because of its alkalinity, calcium hydroxide had increasingly deleterious effects on the fermentation rate and loaf volume when absorption water containing 100, 500, and 1,000 p.p.m. was used to disperse the yeast, as is normally the practice in dough preparation. When the yeast was mixed with the dry ingredients before the calcium hydroxide solution was metered into the mixer, the adverse effect of the alkali was not observed. Evidently the buffering effect of flour, which is considerable, was sufficient to bring even the 1,000 p.p.m. solution down to a pH that was not inhibitory to the yeast. Results obtained with magnesium oxide solutions were not as extreme as those with calcium oxide.

Haas (1927) believed that few salts normally occur in water in sufficient concentrations to exert an important effect on dough fermentation. He recognized, however, that the alkali chlorides and sulfates, and particularly calcium sulfate, were beneficial to doughs because of their gluten strengthening function.

Chlorides and fluorides from the absorption water are thought to have little effect in bread doughs (Kirk 1951; Dunn 1941). Cryns (1936) indicated that his experience showed the use of chlorinated water shortened fermentation time, at least to a slight extent, and improved bread quality, especially if lower grade or weaker flours were used.

In many instances, the adverse effects of water of poor quality may be offset by adding appropriate amounts of certain types of dough improvers. Kirk (1951) published a summary of this type of treatment which is reproduced in Table 21.

It is difficult to explain why waters differing by 100 to 200 p.p.m. can have detectable effects on dough properties when the ionic contribution

TABLE 21

RELATIONSHIP OF WATER CHARACTERISTICS TO YEAST FOOD REQUIREMENTS[1]

Type of Water	Sub-class of Water	Type of Yeast Food Needed	Amount of Yeast Food Needed	Other Special Treatment
I. Acid (Below 7 pH)	A. Soft (less than 120 p.p.m.)	Regular	Normal	Salt in sponge (plus calcium sulphate in extreme cases)
	B. Moderately hard (120–180 p.p.m.)	Regular	Normal	None
	C. Hard (more than 180 p.p.m.)	Regular	Less than normal	Malt in sponge (in extreme cases)
II. Average (7 to 8 pH)	A. Soft	Regular	More than normal	None
	B. Moderately hard	Regular	Normal	None
	C. Hard	Regular	Less than normal	Malt in sponge
III. Alkaline (Above 8 pH)	A. Soft	Acid or regular plus calcium acid phosphate	More than normal	Calcium acid phosphate (in extreme cases)
	B. Moderately hard	Acid	Normal	None
	C. Hard	Acid	Less than normal	Much malt (plus vinegar or lactic acid in extreme cases)

[1] Kirk (1951).

by other ingredients far outweighs the contribution of the water minerals. Skovholt (1948) suggested that the trace elements found in water might play a more important role than ordinarily believed. It is known that less than one part per million of vanadium will substantially modify dough characteristics. Cadmium is also very effective, the amount of this element entering a dough from plated mixer arms being sufficient to retard fermentation. Both of these substances apparently act by suppressing yeast metabolism.

OTHER PRODUCTS

The literature on the influence of minor constituents of ingredient water on the quality of many other important foodstuffs is surprisingly fragmentary. In this short section will be included brief discussions of the current knowledge of the interaction of natural waters and several different food products.

Rice cooked in alkaline water such as most natural hard water, softened water, and water from many streams, usually has a yellow hue, while the same rice cooked in distilled water would have a snowy white appearance. Cauliflower, white cabbage, some potatoes, and particularly white onions show the same reaction. Lowe (1955) attributes this effect to hydrolysis of the flavones from their normal glucoside combinations.

Alkaline water will sometimes give a bluish tinge to red fruit juices, depending on the alkalinity of the water and the acidity, showing that the anthocyanin pigment involved is pH sensitive.

It has been claimed that better color is obtained in corned beef and other pickled meats when they are washed in softened water. The mechanism is obscure but may involve the reaction of calcium ions with reactive sites created on the pigment molecules by the reducing agents used in pickling.

A bitter taste can be imparted to cucumber pickles by calcium and magnesium ions from the brine. An undesirable toughness may also result from the previously described pectin-calcium reaction, and this is a textural defect readily distinguishable from the customary crispness. Iron will darken the pickles by interaction with traces of tannin present in the spices and vinegar.

When citrus fruits are washed in hard water, a film forms on the surface, dulling the appearance of the fruit. Dust and dirt tend to adhere to the film, further decreasing the natural sheen of the product.

BIBLIOGRAPHY

ANON. 1934. Water supply in the canning factory. Canner Food 3, 288–290.

BASHFORD, T. E. 1947. Infected cooling water and its effect on spoilage in canned foods. Proc. Soc. Applied Bacteriol. 1947, 46–49.

BIGELOW, W. D., and STEVENSON, A. E. 1923. The effect of hard water in canning vegetables. Natl. Canners Assoc. Research Lab. Bull. No. 20–L.

BONNET, J. 1955. Water in the agricultural industries. Bull. centre belge étude et document. eaux (Liege) No. 28, 70–80.

BORUFF, C. S., SMITH, B., and WALKER, M. G. 1943. Water for grain alcohol distilleries. Ind. Eng. Chem. 35, 1211–1213.

BROOKE, M. M. 1939. Survey of basic work on the effects of various minerals on fermentation. Proc. Am. Soc. Bakery Engineers 1939, 84–86.

BROWN, E. B. 1939. The effect of minerals of water on baking, using sponge and straight doughs. Proc. Am. Soc. Bakery Engineers 1939, 88–91.

COLE, B. C. 1949. Water treatment spells better beverages. Food Inds. 21, 1056–1058.

CRYNS, J. 1936. Water—five questions. Proc. Am. Soc. Bakery Engineers 1936, 77–78.

DE CLERCK, J. 1957. A Textbook of Brewing. Chapman and Hill, London.

DUNN, J. A. 1941. Salt and water. Proc. Am. Soc. Bakery Engineers 1941, 70–71.

EHRENFELD, C. H., and GIBBS, R. E. 1928. Water for Icemaking and Refrigeration. Nickerson and Collins, Chicago.

HAAS, L. W. 1927. Water in baking. Proc. Am. Soc. Bakery Engineers 1927, 80–81.

HARRIS, J. O. 1959. The use of water in the brewing industry. Proc. Soc. Water Treatment and Examination 8, 138–140.

HERON, H. 1935. The mineral constituents of brewing waters and their influence in brewing operation. J. Inst. Brewing 41, 283–296.

HLYNKA, I., and CHANIN, W. G. 1957. Effect of pH on bromated and unbromated doughs. Cereal Chem. 34, 371–377.

HOAK, R. D. 1956. Industrial water requirements. Public Works 87, No. 11, 160.

HUENINK, L., and BARTOW, E. 1915. The effect of the mineral content of water on canned foods. Ind. Eng. Chem. 7, 495–496.

INGLETON, J. F. 1962. The importance of water in confectionery manufacture. Confectionery Prod. 28, 1065–1068.

JORDAN, S. 1930. Confectionery Problems. National Confectioners Association, Chicago.

JUVRUD, I. O. 1939. Effect of minerals of water on baking, using different yeast foods. Proc. Am. Soc. Bakery Engineers, 1939, 97–101.

KASTNER, J. 1955. The importance of organic matter and products of its decomposition in brewing waters. Water Pollution Abstr. 28, 1074.

KIRK, D. J. 1951. Effects of hardness and acidity of water on fermentation. Bakers Digest 25, No. 6, 28–30, 34.

KNORR, F. 1953. Biology of brewing water. Brauer u. Malzer 6, No. 2, 8.

KOHMANN, E. F. 1923. The effect on canned foods of industrial wastes in the water supply. Natl. Canners Assoc. Circ. No. 4–L.

LAMMERS, F. J. 1934. Water purification in the modern brewery and distillery. Ind. Eng. Chem. 26, 1133–1138.

LANCEFIELD, S. 1936. Water for the canning factory. Food 5, 353–356.

LECLERC, J. 1948. Brewery water and its analysis. Chimie et Industrie 59, 484–490.

LEE, F. A., and WHITCOMBE, J. 1945. Blanching of vegetables for freezing. Effects of different types of potable water on nutrients of peas and snap beans. J. Food Res. 10, 465–468.

LOVING, H. J. 1939. Effect of minerals of water on baking, using different yeasts. Proc. Am. Soc. Bakery Engineers 1939, 92–97.

LOWE, B. 1955. Experimental Cookery from the Chemical and Physical Standpoint. 4th Edition. John Wiley and Sons, New York.

MASELLI, J. A. 1960. Water. In Bakery Technology and Engineering. Edited by S. A. Matz. Avi Publishing Co., Westport, Conn.

MASTERS, H. 1918. An investigation of the methods employed for cooking vegetables, with special reference to the losses incurred. I. Dried legumes. Biochem. J. 12, 231–247.

MASTERS, H., and GARBUTT, P. 1920. An investigation of the methods employed for cooking vegetables, with special reference to the losses incurred. II. Green vegetables. Biochem. J. 14, 75–90.

MCKEE, J. E., and WOLF, H. W. 1963. Water Quality Criteria. 2nd Edition. State Water Quality Control Board, Sacramento, Calif.

Moore, A. V., and Kapsalis, J. G. 1963. The effect of water minerals on certain properties of reconstituted nonfat dry milk. Food Technol. *17,* 920–922.

Mott, K. 1940. The water quality requirements of the soft drink industry. J. Am. Water Works Assoc. *32,* 1512–1516.

Ohlmeyer, D. W. 1959. Brewing. *In* The Chemistry and Technology of Cereals as Food and Feed. Edited by S. A. Matz. Avi Publishing Co., Westport, Conn.

Pozen, M. A. 1940. Water in the brewery. Modern Brewery Age *23,* No. 3, 67.

Rahlén, E. 1931. A guide to the choice of water for dairy use, particularly with regard to the requirements of cheese dairies. Medd. Kgl. Lantbruksstyrelsen *285,* 5–13.

Raux, J. 1940. Water. Brasseur franc. *4,* No. 7, 140.

Siebel, F. P., Jr. 1928. The use of water in baking. J. Am. Water Works Assoc. *19,* 337–340.

Skovholt, O. 1948. Water in baking. Bakers Digest *22,* No. 4, 65–66, 81.

Steffens, W. 1929. Water in the manufacture of ice. Vom Wasser *3,* 118–123.

Toyoda, K., Kamatsu, H. and Go, E. 1961. Investigation of water for ice manufacturing. I. Analysis of ice-manufacturing in Tokyo and its vicinity and distribution of salts in frozen ice. J. Am. Water Works Assoc. *53,* No. 3, P & R Section, 80.

Wayt, J. A. 1936. Discussion on "What is alkaline water and what does it do to bread dough?" Proc. Am. Soc. Bakery Engineers *1936,* 70–72.

Webber, H. F. P., and Taylor, L. 1954. Effect of fluorine on fermentation. Brewers Digest *29,* No. 12, 59–61.

Weckel, K. G. 1942. Watch your water supply—it affects food quality. Food Inds. *14,* No. 1, 47–50, 85.

West, P. W. 1944. The ice industry. Chem. Eng. News *22,* 718.

CHAPTER 14

Behavior of Water in Freezing
and Frozen Storage

INTRODUCTION

Any substantial cooling of a water-containing foodstuff will have an effect on its textural qualities. The viscosity of the aqueous phase is increased. Fats and gels become firmer. Poorly soluble substances may precipitate and become perceptible as granular inclusions. Solubility of gases will increase with possible increases in the density of the product as bubbles shrink. However, the most significant changes and those of most interest to the food technologist are the qualitative alterations accompanying the solidification of the water, i.e., the freezing of the substance. The behavior of water during the freezing stage and during frozen storage can markedly influence the characteristics of a foodstuff, especially the texture.

When water solidifies, it ordinarily assumes a crystalline habit. Dorsey (1940) summarized the then current opinions as to the types of ice which can exist. Ordinary ice, Ice-I, is crystalline and has a conchoidal fracture and a vitreous luster. It melts at 32°F. at one atmosphere pressure and is the only variety of ice which is lighter than water under the same conditions of temperature and pressure. There are indications that Ice-I can exist as α-ice which is hexagonal and appears when water near 32°F. solidifies, and β-ice which is rhombohedral and appears when water that is supercooled by at least a few degrees freezes. The ordinary form of ice is the hexagonal or tridymite type. Ice-II is characterized by a side-centered orthorhombic cell containing eight molecules. Ice-III is a body-centered orthorhombic cell containing 16 molecules and having a:b:c = 1.73:1:1.22.

Under special conditions water can assume a vitreous or glass-like state when it solidifies. All investigators agree that the production of a completely vitreous mass of water is quite difficult. It appears to be stable only at very low temperatures. Burton and Oliver (1936) condensed water vapor on a chilled surface and examined the deposit by X-ray diffraction to determine the presence or absence of crystallinity. Below —166°F. the condensate was vitreous, while above this temperature, a semicrystalline or crystalline solid was obtained. The transformation temperature is nearer —202°F. according to the later work of Pryde and Jones (1952), who also found that specimens of glassy water containing

189

traces of crystalline material revert completely to the crystalline form when the temperature is raised to —200°F. Tammann and Büchner (1935) indicated that the formation of glass-ice can be facilitated by the addition of certain compounds that are effective in reducing the crystallization velocity. There would seem to be some question as to whether the glass-ice obtained in this manner is truly amorphous or merely composed of crystals of submicroscopic size, thus superficially resembling the amorphous state.

An essential feature of ice crystals is that they are held together by hydrogen bonds exclusively. X-ray diffraction patterns of ice crystals indicate that each oxygen atom is surrounded by four nearly equidistant hydrogen atoms tetrahedrally arranged, and these in turn are bridged to other oxygen atoms thus building up a symmetrical structure. The distance between oxygen atoms is close to 2.76 Å. The most precise measurements available seem to indicate that the mirror-symmetric bond (along the principal axis) is about 0.01 Å shorter than the central symmetric bond (at 32°F.). At lower temperatures, these bond lengths tend to equalize at a value about one per cent lower than at 32°F.

When ice crystals grow in solutions, rather than in pure water, the gross structures can assume many different shapes, and these have been studied extensively by several investigators. Hallett and Mason (1958) grew ice crystals on a fiber suspended in a water-vapor diffusion cloud chamber in which the temperature and supersaturation of the environment could be varied independently. The crystals underwent the following transitions of habit in the temperature range 32° to —22°F., plates→ needles→ hollow prismatic columns→ plates→ dendritic crystals→ plates→ prismatic columns. Changes were controlled mainly by the temperature of the environment while large variations of supersaturation influenced only secondary features such as the development of dendritic forms. Atmospheric aerosols had little or no effect on the shapes. Crystals of composite habit were made by changing conditions midway in their growth. Small quantities of organic vapors such as camphor and isobutyl alcohol modified the habit to an extent dependent upon their concentration.

Luyet (1960) and his co-workers have published many papers dealing with the forms of ice crystals which develop in solutions. He has photographed changes in droplets of aqueous solution held between two cover glasses and plunged into baths held at 32° to —150°F. He classified the types of ice formation thus obtained into nine categories based on their morphology, as shown in Table 22. The rate of freezing determines whether spheroidal masses or the expected hexagonal forms appear. The former seem to have a relatively coarse arrangement of an indefinite number of component elements radiating from centers of crystallization.

TABLE 22

STRUCTURE OF SOLIDIFIED MATERIAL AND TYPES OF CRYSTALLINE
FORMATIONS OBSERVED IN THE FREEZING OF AQUEOUS SOLUTIONS[1]

Structure	Arrangement	Form
Crystalline	Hexagonal	1. Prisms 2. Stars 3. Rosettes 4. Skeletons 5. Transition forms
	Radial	6. Spherulites (plain) 7. Spherulites (evanescent) 8. Lobed formations (irregular rosettes)
	Diffuse	9. Recrystallization clouds
Amorphous		

[1] Luyet (1960).

SUPERCOOLING

The melting temperature of a particular aqueous system is usually sharp and easily reproducible. It has frequently been observed, however, that pure water and many aqueous systems can be cooled considerably below the normal melting point without the appearance of ice, i.e., they can be supercooled. The initial appearance of crystals in these supercooled liquids is followed immediately by solidification of the whole mass. Eutectics can supercool in the presence of ice crystals, i.e., ice crystals do not provide nuclei for the crystallization of eutectics. The initiation of the phenomenon of crystallization apparently depends upon the presence of suitable molecular arrangements within the liquid.

The formation of foci around which crystals can start to grow is called nucleation, and the factors governing it have been subjected to considerable study not only in the case of aqueous solutions and pure water but also for molten metals and salts. It has been found that many of the principles are universally applicable. In aqueous systems, there appear to be two mechanisms which may function to provide foci for crystal growth. These are homogeneous nucleation and heterogeneous nucleation.

Nucleation

Homogeneous nucleation results from random local fluctuations in the density and configuration of pure water leading to the chance spontaneous occurrence of regions having crystal-type orientation and dimensions.

According to Turnbull's theory of homogeneous nucleation (Turnbull et al. 1948), the equilibrium number of nuclei of radius r in a supercooled liquid at an absolute temperature, T, is proportional to $\exp[-\Delta F^*/kT]$, in which the free energy change in the formation of the nucleus is $\Delta F^* = -\frac{4}{3}\pi r^3 \Delta F_v + 4\pi r^2 \sigma$. In this equation, σ is the ice:water interfacial tension, ΔF_v is the bulk free energy change in the transformation of water to ice and is given approximately by $\Delta F_v \cong L_f \Delta T/T$, where L_f is the latent heat of fusion at equilibrium temperature, and $\Delta T = T_0 - T$ when T_0 is the equilibrium temperature. A critical nucleus is a nucleus of radius such that $d(\Delta F^*)\ dr = 0$. Upon differentiation it is found that for the critical radius, r_o, and the associated free energy change, $r_c = 2\sigma/\Delta F_v$, the equation $\Delta F^* = (16\pi/3)\sigma^3/\Delta F_v^2$ applies. When the numerical values $\sigma = 32.8$ ergs/cm.^{2} and $\Delta F_v = 1.23 \times 10^7 \Delta T$ ergs/cm.^{3} are subtracted, a value of $r_o = 532/(T_0 - T)$ is obtained for the value of a critical radius in Å at temperature T. From this, it can be concluded that the size of a critical nucleus for super-cooling to $-14°$F. is about 50 Å, and at $-58°$F. is about 5 Å. The number of molecules of water in a critical nucleus would then be $n_c = 2.1 \times 10^7/(\Delta T)^3$.

The critical nucleation temperature (homogeneous) for water appears to be about $-36°$F. It has been shown that small suspended water droplets invariably change to snow at this temperature while the droplets may persist for long times in the liquid form at a slightly higher temperature. From the calculations of Fisher *et al.* (1949), it can be seen that the probability of nucleation becomes extremely high near $-36°$F. It follows that bulk pure water cannot be supercooled below about $-36°$F. On the other hand, homogeneous nucleation near 32°F. has low probability, especially in a small volume.

Dorsey (1948) found that the temperatures of spontaneous freezing for different samples of nominally pure water tested by him ranged from 27°F. to $-4°$F. No doubt heterogeneous nucleation (sometimes called catalytic nucleation) came into play on these occasions. This type of nucleation is the result of the deposition of water molecules in crystalline array on some foreign particle. In some way which is not well understood, the non-water suspensoids provide a foundation on which the water molecules can build the particular ordered arrangement characteristic of a crystal. Perhaps on the relatively enormous surface of the suspended particle there occurs occasionally an arrangement of atoms which, by chance, has the exact dimensions and charge of a region on the surface of a water crystal. Relatively small volumes of solutions, or the order of ten milliliters, can be supercooled without difficulty, and the extent to which a given sample can be supercooled is often reproducible to a fraction of a degree.

Langham and Mason (1958) reporting a study on the heterogeneous and homogeneous nucleation of supercooled water, showed there is a proportionality between the log of droplet dimensions (i.e., volume) of distilled water, in the range of ten microns to one centimeter in diameter, and its freezing temperature for a constant rate of cooling. Such a relationship appears to characterize the heterogeneous nucleation of water droplets containing small foreign nuclei and can be interpreted in terms of the droplets being "infected" with atmospheric aerosols whose ice-nucleating ability is known to increase roughly logarithmically with decreasing temperature.

Turnbull (1950A) calculated the interfacial energies, σ, between crystal nuclei and the corresponding liquids from the nucleation frequencies of small droplets on the basis of the theory of homogeneous nucleation. Energies of interfaces, σ_y, one atom thick and containing N atoms were calculated from the σ's. The ratio of σ to the gram atomic heat of fusion, ΔH_f, was about 0.32 for water. For most metals, the figure was nearer 0.45.

Stephenson (1960) did not consider Turnbull's theory strictly applicable to heterogeneous nucleation in hydrated colloids. The growth of ice crystals in the hydration water of the colloid is regarded as altering both the bulk and surface energy terms, and ΔF_v and σ in Turnbull's equations will refer to the change in bulk free energy and the interfacial tension between "bound" water and ice. If the net effect is to decrease the free energy required for the formation of a nucleus of given size in the hydration layer, nuclei up to the approximate size of the hydrated colloid will be formed with increasing frequency. These cannot grow further, however, until the supercooling is such that the radius of a critical nucleus becomes less than the radius of a colloidal particle. In view of the inability to super-cool biological materials more than 14°F. below their freezing point reported by Lusena (1955), the relation $r_c = 532\Delta/T$ becomes important, because it yields a critical radius of about 50 Å, well within the range of the radii of hydrated macromolecules and their aggregates.

It appears that a slow rate of heat exchange brings relatively fewer of these sites into action, and a smaller number of large ice crystals will form under these conditions. As the rate of freezing is increased, more of the loci become effective and the number of crystals increase making the final crystals smaller and more uniform in size. When temperatures of about −40°F. can be obtained with sufficient rapidity, a very uniform deposit of many microscopic crystals is formed. This situation is thought to be the optimum result for the freezing of foodstuffs. It also follows from the critical temperature of homogeneous nucleation, that the achievement of temperatures lower than about −40°F. within a product do not provide additional advantages during freezing. Of course, it may be necessary to bring the outer layers of a bulky item to lower temperatures in order to quickly bring the center of it to the critical nucleation temperature.

CRYSTALLIZATION VELOCITY

Once nucleation has been established, the growth of the seed crystals is governed by the rate at which water molecules can diffuse to the surface of the crystal and the rate at which the heat of fusion can be dissipated from its surface. At temperatures a few degrees below 32°F., and lower, growth rates appear to be controlled mostly by the diffusion of water so that the rate of increase in the size of the crystals is reduced exponentially with decreasing temperature (Tammann 1925). This reduction in growth rate allows more nuclei to come into play before all of the water is consumed by the foci originally formed, and is one of the reasons for the appearance of smaller crystals when lower temperatures are used.

Lusena (1955), in a repetition and extension of the observations of Tammann and Büchner (1935), demonstrated that relatively small amounts of certain alcohols, glycols, sugars, and proteins can considerably retard crystallization velocity, some of these reducing the rate by a factor of ten at concentrations of about five per cent. Ethanol reduced the crystallization velocity by as much as $1,000 \times$ at a concentration of 30%. Lusena suggested that these compounds may act as impurities included in the oriented structure of a growing crystal face by virtue of their bound and electrostatically associated water layers or as interfacial hydrates, creating obstacles to the subsequent growth of that particular crystal face. In concentrated solutions of gelatin, sucrose, and glycerol, crystals appeared at the freezing point when slow cooling rates were used but only below the freezing point when fast cooling was used (Lusena and Cook 1954). Increasing either the concentration of solute or the rate of cooling increased the difference between the freezing point and the temperature of freezing.

Those materials which hinder the diffusion of water, and especially hydrocolloids, seem to be effective in reducing the crystal size probably because they reduce the growth of crystals at each foci. In the absence of substances which reduce the diffusion of water to the crystal face, growth from the initial foci can be very rapid and a few seeds are sufficient to inoculate large volumes.

CONCENTRATION OF SOLUTES

As a solution is cooled, solutes may precipitate out before the freezing point is reached because of their decreasing solubility as the temperature is lowered. After the freezing point is reached, those solutes which do not precipitate are increasingly concentrated as the water is taken out of the solution in the form of ice crystals. At a certain temperature, which depends upon the components in the solution, a eutectic mixture will begin to separate.

More than one eutectic point can be observed in a solution containing several solutes. Each particular eutectic will contain constant proportions of water and one or more solutes. The solid phase which separates consists of closely intermingled crystals of ice and crystals of the solutes or of their hydrates. Since each eutectic mixture, as it freezes, takes out more water, the residual solution continues to increase in concentration until the lowest-freezing eutectic solidifies and removes remaining liquid water. These highly concentrated solutions remain in contact with colloidal substances and may irreversibly damage them by causing precipitation or denaturation. The pools of unfrozen concentrated solution will tend to run together and coalesce, where this is possible, forming pockets of salts which may further damage the surrounding structural materials when the product is thawed.

The most favorable situation for preservation of the original condition of colloidal materials appears to be the separation of eutectics in the smallest possible volumes. This, too, is facilitated by using very low freezing temperatures and conditions leading to rapid heat transfer so that eutectic droplets are entrapped between rapidly forming ice crystals, and the eutectics themselves should solidify before they have an opportunity to coalesce into large pools which may cause macroscopic damage to structural elements.

In living systems, or in materials derived from them, there may be present various dissolved low molecular weight organic compounds such as ethanol, acetic acid, and lactic acid, which form eutectic mixtures with low freezing points. Salts will also be present and will participate in the formation of eutectic mixtures. In manufactured foods, especially those containing only a few components, the freezing point may be fairly sharp, with perhaps only one eutectic point occurring. Animal and vegetable substances, however, may have a continuous succession of eutectics over a wide range of temperature so that, instead of one temperature plateau, a gradual curve composed of several eutectic points blending into one another is observed.

Gases present in dissolved form are also concentrated by the freezing out of pure water and ultimately may be thrown out of solution as their limit of solubility in the remaining liquid is reached.

Even in supposedly pure ice, there is between the crystals a material with a lower melting point than the crystals themselves. When this material is liquid, it dissolves the surface of the ice crystals it contacts even though the temperature is below the melting point of the crystal. This boundary material may be, in some cases, less than eight microns thick (Dorsey 1940).

MIGRATORY RECRYSTALLIZATION

In a system containing ice crystals of different sizes in contact with one another through either an aqueous solution or the vapor state, the larger crystals will tend to grow at the expense of the smaller ones. This "migratory recrystallization" is one manifestation of a more general mechanism of widespread importance operating also in saturated salt solutions and in foams. The proximate cause is the larger surface energy of the smaller objects. The rates of growth and decrease are greatest in the microscopic size range of crystals. As the crystals become larger, the differences in surface energies become relatively much less. Furthermore, the velocity of change is greater at higher temperatures (but below the freezing point, of course).

The change of crystal size is of considerable importance in maintaining product quality during frozen storage. The crystals initially deposited should be small and uniform in size in order to retard the formation of large crystals which may mechanically damage the tissues.

Low temperatures reduce crystal growth by decreasing the vapor pressure and the diffusion of water molecules. Rey (1957) microscopically observed that migratory recrystallization is rather slow up to 14°F. and that up to this temperature crystals will remain small in size during long intervals of time. Luyet (1957B) observed and photographed migratory recrystallization on microscope slides when preparations containing small ice crystals were either maintained for some time at a given low temperature or were warmed up gradually.

An early, but important, paper by Thomson (1861) showed that migratory recrystallization must occur whenever it is possible for molecules to pass by any means from a stressed crystal to one that is less stressed. Some obvious means of passing between crystals are by fusion and resolidification or sublimation and resolidification. Thomson stated "I came also to the more general inference that stress tending to change the form of any crystals in the saturated solution from which they have been crystallized must give them a tendency to dissolve away, and to generate, in substitution for themselves, other crystals free from the applied stresses or any equivalent stresses."

FREEZING OF BIOLOGICAL SYSTEMS

Freezing of cells in a liquid environment results first in the formation of extracellular crystal nuclei (Chambers and Hale 1932). During their subsequent growth, these collect water both from the extracellular space and from within the cell until all of the available water has been incorporated into ice crystals. The cell membrane forms a barrier to crystal penetration in living and in many dead cells. However, the growing crystals can mechanically disrupt the structures surrounding them. This has been reported by many workers. Luyet (1960) made microscope sections of ice forms frozen in gelatin (e.g., in 40% gelatin gel frozen at 5°F.) and found that growing ice columns had tunneled their way through the gelatin. The water content of the gel was noticeably reduced at a distance from the crystal of from 20,000 to 30,000 times the diameter of a water molecule.

In higher plant and animal cells, slow cooling allows intracellular water to leave a cell and freeze externally, thereby reducing the probability of intracellular freezing (Siminovitch and Scarth 1938). However, this tendency is overcome by freezing at very low temperatures and intracellular crystal growth then occurs more or less simultaneously with extracellular freezing. Thus it may be that the lack of appearance of visible crystals inside the cell is due more to the restrictions on growth rather than the absence of nucleation, the enlargement of the extracellular forms being at the expense of internal crystals which are too small to be seen. Luyet and Gibbs (1937) observed that different cells in a single microscopic preparation congealed separately and at different times. They also stated "The changes in opacity during freezing indicate that numerous submicroscopic crystals are formed in the cells at the time of flashing and that these crystals increase considerably in size during the first seconds after their formation and continue to grow, less and less, at least for a few seconds."

A quotation from Meryman (1957A) is pertinent:
"Possibly the single most important concept in biological freezing as well as the most difficult for the biologist to assimilate is the following: despite the collective and individual complexity of living cells, freezing in a biological matrix represents nothing more than the removal of all available water and its isolation into inert foreign bodies, the ice crystals. . . . The transfer of water out of solution into ice crystals is a very straightforward phenomenon and is, in effect, simply dehydration with the water sequestered locally in the tissue rather than removed. . . . From this dual event—dehydration with foreign body formation—stem all of the physiological and biochemical events subsequent to freezing."

Perhaps 5 to 10% of the water content (as determined by a drying procedure) in a tissue is "bound" to other cellular constituents by forces which preclude it from entering into the formation of ice or eutectics.

Studies which have been made with living tissues are often instructive and pertinent to food problems because the highest degree of structural and biochemical integrity is required for maintaining the living state. In other words, conditions which would lead to the continuance of viability or physiological normality in a tissue would undoubtedly also be conducive to the preservation of the original organoleptic properties of a foodstuff. Some of these studies are briefly described below.

Many authors have confirmed observations that glycerol when added to blood cells, spermatozoa, etc., before freezing, leads to improved retention of original characteristics. Lovelock (1954) found that sucrose provided little or no protection to red blood cells during freezing. He concluded that the highly hypertonic solutions which develop in between the initial freezing of living cells and the eutectic points are responsible for the major part of the alterations due to freezing, and that thermal shock is, in part, osmotic shock. On the other hand, Polge and Soltys (1960) reported that small molecular concentrations of xylose, sucrose, or glucose provided a relatively high degree of protection to trypanosomes and bull spermatozoa. The latter authors concluded that one of the features differentiating their results from those of Lovelock was the absence of added salts in their experiments. Luyet and Grell (1936) found that cellular constituents of onion root tips were more readily stratified (usually into two layers) by the ultracentrifuge after freezing than before, suggesting a more fluid condition such as might result from breakdown of the gels or a kind of syneresis.

Christian and Ingram (1959) found that the freezing points of bacterial cells were always just slightly less than the freezing points of the media on which they were grown. This was true for media having freezing points of 25°F. to about 3°F.

FREEZING OF FOODS

As a food processing operation, freezing is generally conducted for one of two reasons: (1) the promotion of desired texture changes, as in the production of ice cream; or (2) the improvement of storage stability.

Ice cream and similar frozen confectionery depend for their texture to a large extent on the size and size distribution of the ice crystals and the completeness of the freezing process, i.e., the proportion of liquid left unfrozen. Other factors contributing to texture are the size and size distribution of lactose crystals, the amount of air whipped into the product, and the amount of fat present. If the rate of freezing is not sufficiently rapid, large ice crystals may develop. Probably the preferred situation is the formation of a crop of very small and uniform ice crystals at an early stage in the freezing process so that the inevitable growth occurring in the subsequent hardening and storage stages does not lead to graininess.

When ice cream, sherbets, and ices are stored for several months they usually become grainy because large ice crystals develop. Fluctuations of temperature or relatively high storage temperatures will tend to accelerate crystal growth (Tressler and Evers 1957). Graininess or sandiness can also result from the development of large lactose crystals. Removal of the lactose from its supersaturated solution in the form of minute crystals during the freezing process reduces the danger of sandiness developing during normal storage. According to Nickerson (1954), the crystals initially formed should not be larger than ten microns, presumably in their largest dimension. Large lactose crystals can develop during storage of ice cream because of inadequate numbers of foci on which the supersaturated lactose can crystallize. Lactose sandiness can be controlled by adjustment of the freezing conditions, or, as suggested by Nickerson, through seeding the mix with lactose crystals.

The addition of "stabilizers" to frozen desert items is a common practice. All of these agents are hydrocolloids such as gelatin and are frequently plant gums and mucilages such as carrageenin. Their mode of function is not universally agreed upon, but it seems likely that they exert their effect by reducing the crystal size of ice, and, probably, of lactose. This would be accomplished by the restrictions the stabilizers exert on water mobility during freezing so that large numbers of nuclei of crystallization would be able to come into operation before freezing is complete. An increase in the rate of freezing would have a somewhat similar effect, as would decreases in the amount of agitation or convection within the liquid. Storage at the lowest possible temperature is indicated as a means of reducing the translocation of water through the lowering of its vapor pressure.

The uniformity of the crystals is more important as an index of texture than is their average size. A relatively few large crystals can create the impression of sandiness even though the great majority of the ice crystals are below perceptible size. Furthermore, as previously stated, the presence of a relatively few large crystals in a system that also contains very many crystals of much smaller size is a condition that contributes to rapid growth of the large crystals at the expense of the smaller ones.

One reason for the effectiveness of the freezing process in promoting storage stability is that it removes water from participation in chemical reactions. For maximum effectiveness, substantially all of the water must be present as ice since partial freezing may result only in a concentration of reactants. This state requires temperatures much below the freezing point of pure water, for most biological materials. Rey (1960) places the eutectic points of whole animal tissues in the neighborhood of $-56°F.$, much lower than usually thought feasible for the frozen storage of foods. There can be a very substantial reduction in spoilage reactions even though some liquid water remains, but the difference is not of the same order as that which occurs when freezing is complete.

It is reasonably certain that enzymatic and microbiological reactions in foodstuffs come to a complete standstill when all of the water is converted to ice. Deteriorative chemical changes such as the development of oxidative rancidity (of course, enzymes can also mediate the development of rancidity) and physical changes such as the growth of ice crystals or loss of water through sublimation, as well as strictly mechanical deterioration through breakage and the like, can continue in the absence of liquid water.

There are disadvantages as well as advantages to the freezing of foodstuffs. Not all of the reactions which occur are desirable. The principal objectionable changes resulting from freezing affect principally the textural characteristics and the appearance of the products. Nutritive value, taste, and odor usually prove to be refractory to the damaging reactions. Some of the undesirable changes which occur are: (1) dehydration as the result of the removal of water from hydrocolloids into ice crystals with the resultant collapse of gels, etc.; and (2) mechanical damage to structural components as the result of the formation and growth of ice crystals.

The size of the ice crystals formed during the freezing process is generally recognized to be related to the extent of textural damage observed in a foodstuff after it is thawed. Slow freezing tends to yield large crystals while fast freezing usually results in the formation of relatively small crystals. The rate of freezing may be considered to be, for most practical purposes, a function of the temperature of the fluid (gas or liquid) by which the object is surrounded or of the solid with which it is in contact, although it is certainly true that the rate of heat transfer is the fundamental controlling process and this depends upon factors other than the temperature differential. However, there are conditions not related directly to the rate of heat transfer which have an effect on the textural damage undergone during freezing. For example, the number of ice crystal foci and their distribution within the foodstuff are influential in this regard. The ease with which water molecules can diffuse to crystallizing sites is important, this being controlled not so much by the extent of binding (in the molecular sense) as by the actual physical discontinuities in the structures, such as air pockets, fatty deposits and other relatively dehydrated areas, pools of eutectics, and, perhaps, cell walls.

Foci of crystallization are doubtless not distributed uniformly in any ordinary foodstuff, so that the opportunity to grow is limited by varying distances. In the case of foci at uniform distances from each other, crystals of similar size will tend to form, and growth of the crystals during storage will occur less readily in such instances than it will in foodstuffs containing crystals of a wide range of sizes. Growth will be most rapid when the bulk of the water is in the form of very small crystals with a generous sprinkling of substantially larger crystals.

When most of the crystals formed during the freezing process are relatively large, the net translocation of moisture between them is slow. The surface energies of large crystals do not vary sufficiently over a wide range of masses to pump water molecules at an appreciable rate. In such cases, the amount of water represented by the small crystals is insufficient to cause much growth even though all of it were transferred to the larger crystals.

The degree of mobility of the molecules is also a factor in the rate of crystal growth during storage. The mobility depends in large part upon the temperature since movement occurs through sublimation so that physical discontinuities and chemical binding do not exert quite as large an influence as they do when water is in the liquid form.

SPECIAL PROBLEMS IN FOOD FREEZING

The freezing of oil-in-water emulsions frequently leads to the breaking of the emulsion with separation of the components when thawed. The aqueous phase which forms the continuous network stabilizing the emulsion is brought from a film-like extended condition to a more condensed crystalline form so that it no longer envelops the fatty constituent. The latter then coalesces and separates, either immediately or upon thawing. Water-in-oil emulsions may or may not be broken by freezing, depending upon the type of oil, size of the water globules, percentage of oil, and, probably, other conditions which have not been fully elucidated.

Most sauces, gravies, etc., break down upon freezing, with curdling or other indications of separation developing when the material is thawed. This is due principally to a dehydration of the stabilizing material. In the case of starch, the gelatinized material is partially retrograded and precipitated. Certain types of starches appear to be less sensitive to these damaging actions, and a considerable amount of study has gone into the selection of starches, natural and modified, which will hold up under the stresses of freezing, frozen storage, and thawing.

Sublimation of ice crystals in the surface layers of foods with loss of their water to the atmosphere, is the cause of freezer burn, a discoloration of frozen foods stored for lengthy periods. The defect is principally in the appearance. Desiccation of surface cells can be very effectively prevented by inclosing the object in a wrap (preferably tight-fitting) which will reduce moisture vapor transfer. The packaging of eviscerated poultry in a shrunk-on Cryovac film has been very successful. Coating the object with a fairly thick glaze of ice or frozen brine accomplishes the same result and is the method used for retailer packs of shrimp as well as some large cuts of meat destined for relatively lengthy cold storage. Kaess and Weidemann (1962) also found that the dipping of slices of liver in relatively concentrated aqueous solutions of polyols, hexoses, sodium chloride, or urea prevented, or greatly reduced, freezer burn in this product. Whether this effect is due to diffusion of the solute into the surface layers of cells or to the diffusion of water from them is not clear.

BIBLIOGRAPHY

BELEHRADEK, J. 1935. Temperature and Living Matter. Protoplasma Monograph 8. Borntraeger Bros., Berlin.

BERNAL, J. D. 1958. A discussion on the physics of water and ice. Proc. Roy. Soc. London 247A, 536–528.

BERNAL, J. D., and FOWLER, R. H. 1933. Pseudocrystalline structure of water. Trans. Faraday Soc. 29, 1049–1956.

BRAGG, W. H. 1922. The crystal structure of ice. Proc. Phys. Soc. London 34, 98–103.

BURTON, E. F., and OLIVER, W. F. 1936. The crystal structure of ice at low temperature. Proc. Roy. Soc. London 153A, 166–172.

BUSWELL, A. M. 1959. Fifty years of advancement in the knowledge of water substance. J. Am. Water Works Assoc. 51, 879–884.

CARLSON, W. A., ZEIGENFUSS, E. M., and OVERTON, J. D. 1962. Compatibility and manipulation of guar gum. Food Technol. 16, No. 10, 50–54.

CHAMBERS, R., and HALE, H. P. 1932. The formation of ice in protoplasm. Proc. Roy. Soc. London 110B, 336–352.

CHRISTIAN, J. H. B., and INGRAM, M. 1959. The freezing points of bacterial cells in relation to halophilism. J. Gen. Microbiol. 20, 27–31.

DAUGHTERS, M. R., and GLENN, D. S. 1946. Role of water in freezing foods. Refrig. Eng. 32, 137–140.

DORSEY, N. E. 1940. Properties of Ordinary Water Substance. Reinhold Publishing Corp., New York.

EIGEN, M., and DE MAEYER, L. 1958. Self-dissociation and protonic charge transport in water and ice. Proc. Roy. Soc. London 247A, 505–533.

ELLIOTT, R. P., and MICHENER, H. D. 1960. Review of the microbiology of frozen foods. Rept. Conference on Frozen Food Quality 1960, 40–61. Western Regional Research Center, Albany, Calif.

FISHER, J. C., HOLLOMON, J. H., and TURNBULL, D. 1949. Rate of nucleation of solid particles in a subcooled liquid. Science 109, 168–169.

FRANK, H. S. 1958. Covalency in the hydrogen bond and the properties of water and ice. Proc. Roy. Soc. London 247A, 481–492.

GOLDBLITH, S. A., KAREL, M., and LUSK, G. 1963A. The role of food science and technology in the freeze dehydration of foods. I. Food Technol. 17, 139–144.

GOLDBLITH, S. A., KAREL, M., and LUSK, G. 1963B. Role of food science and technology in the freeze dehydration of food. II. Food Technol. 17, 258–263.

GORTNER, R. A. 1949. Outlines of Biochemistry. Third Edition. John Wiley and Sons, New York.

GRÄNICHER, H. 1958. Lattice disorder and the physical properties connected with the hydrogen arrangement in ice crystals. Proc. Roy. Soc. London 247A, 453–461.

GREAVES, R. I. N. 1962. Minimizing product damage when drying biological solutions. *In* Freeze-drying of Foods. Edited by F. R. Fisher. National Academy of Sciences, Washington.

HALE, W. S., and COLE, E. W. 1963. A freezing technique for concentrating pre-ferments. Cereal Chem. *40*, 287–290.

HALLETT, J., and MASON, B. J. 1958. The influence of temperature and super-saturation on the habit of ice crystals grown from the vapor. Proc. Roy. Soc. London *247A*, 440–453.

HUGGINS, M. L. 1936. Hydrogen bridges in organic compounds. J. Org. Chem. *1*, 407–456.

KAESS, G., and WEIDEMANN, J. F. 1962. Freezer burn as a limiting factor in the storage of animal tissue. IV. Dipping treatments to control freezer burn. Food Technol. *16*, No. 12, 83–86.

KUPRIANOFF, J. 1960. Physical and biochemical changs in frozen foodstuffs. Bull. Int. Inst. Refrig. *40*, 775–790.

KUPRIANOFF, J. 1962. Some factors influencing the reversibility of freeze-drying of foodstuffs. *In* Freeze-drying of Foods. Edited by F. R. Fisher. National Academy of Sciences, Washington.

LANGHAM, E. J., and MASON, B. J. 1958. The heterogeneous and homogeneous nucleation of supercooled water. Proc. Roy. Soc. London *247A*, 493–504.

LIPSETT, S. G., JOHNSON, F. M. G., and MAASS, O. 1927. The surface energy and the heat of solution of solid sodium chloride. J. Am. Chem. Soc. *49*, 925–943.

LIPSETT, S. G., JOHNSON, F. M. G., and MAASS, O. 1928. The surface energy of solid sodium chloride. III. The heat of solution of finely ground sodium chloride. J. Am. Chem. Soc. *50*, 2701–2703.

LIVINGSTON, H. K. 1944. Cross-sectional areas of molecules adsorbed on solid surfaces. J. Am. Chem. Soc. *66*, 569–573.

LONSDALE, K. 1958. The structure of ice. Proc. Roy. Soc. London *247A*, 424–434.

LOVELOCK, J. E. 1954. Biophysical aspects of the freezing of living cells in preservation and transplantation of normal tissues. Ciba Foundation Symposium Chem. Structure Proteins *1954*, 131–140.

LOVELOCK, J. E. 1957. The denaturation of lipid-protein complex as a cause of damage by freezing. Proc. Royal Soc. London *147B*, 427–433.

LUSENA, C. V. 1955. Ice propagation in systems of biological interest. III. Effect of solutes on nucleation and growth of ice crystals. Arch. Biochem. Biophys. *57*, 277–284.

LUSENA, C. V., and COOK, W. H. 1954. Ice propagation in systems of biological interest. II. Effect of solutes at rapid cooling rates. Arch. Biochem. Biophys. *50*, 243–251.

LUYET, B. J. 1957A. An analysis of the notions of cooling and of freezing velocity. Biodynamica 7, 293–336.

LUYET, B. J. 1957B. On the growth of the ice phase in aqueous colloids. Proc. Roy. Soc. London *147B*, 434–451.

LUYET, B. J. 1960. On the mechanism of growth of ice crystals in aqueous solutions and on the effect of rapid cooling in hindering crystallization. Recent Research in Freezing and Drying *1960*, 3–22.

LUYET, B. J. 1962. Recent developments in cryobiology and their significance in the study of freezing and freeze-drying of bacteria. *In* Proc. Low Temperature Microbiology Symposium, Campbell Soup Co., Camden, N. J.

Luyet, B. J., and Gehenio, P. M. 1940. The mechanism of injury and death at low temperatures. A review. Biodynamica 3, 33–99.

Luyet, B. J., and Gibbs, M. C. 1937. On the mechanism of congelation and death in the rapid freezing of epidermal plant cells. Biodynamica 1, No. 25, 1–18.

Luyet, B. J., and Grell, S. M. 1936. A study with the ultracentrifuge of the mechanism of death in frozen cells. Biodynamica 1, No. 23, 1–16.

Luyet, B. J., and Rapatz, G. 1957. An automatically regulated refrigeration system for small laboratory equipment and a microscope cooling stage. Biodynamica 7, 337–345.

Meryman, H. T. 1956. Mechanics of freezing in living cells and tissues. Science 124, 515–521.

Meryman, H. T. 1957A. Physical limitations of the rapid freezing method. Proc. Roy. Soc. London B147, 452–459.

Meryman, H. T. 1957B. Tissue freezing and local cold injury. Physiol. Revs. 37, 233–251.

Meryman, H. T. 1960. The mechanisms of freezing in biological systems. In Recent Research in Freezing and Drying. Edited by A. S. Parkes and A. V. Smith. Blackwell Scientific Publications, London.

Meryman, H. T. 1962. Introductory survey of biophysical and biochemical aspects of freeze-drying. In Freeze-drying of Foods. Edited by F. R. Fisher. National Academy of Sciences, Washington.

Moran, T. 1925. Effect of low temperature on hen eggs. Proc. Royal Soc. London B98, 436–456.

Nei, T. 1960. Effects of freezing and freeze-drying on micro-organisms. In Recent Research in Freezing and Drying. Edited by A. S. Parkes and A. V. Smith. Blackwell Scientific Publications, Oxford.

Nickerson, T. A. 1954. Lactose crystallization in ice cream. I. Control of crystal size by seeding. J. Dairy Sci. 37, 1099–1105.

Nickerson, T. A., and Pangborn, R. M. 1961. The influence of sugar in ice cream. III. Effect on physical properties. Food Technol. 15, 105–106.

Peterson, A. C. 1962. An ecological study of frozen foods. In Proceedings Low Temperature Microbiology Symposium. Campbell Soup Co., Camden, N. J.

Polge, C., and Soltys, M. A. 1960. Protective action of some neutral solutes during the freezing of bull spermatozoa and trypanosomes. In Recent Research on Freezing and Drying. Edited by A. S. Parkes and A. V. Smith. Blackwell Scientific Publications, Oxford.

Pryde, J. A., and Jones, G. O. 1952. Properties of vitreous water. Nature 170, 685–688.

Rapatz, G., and Luyet, B. 1957. Apparatus for cinemicrography during rapid freezing. Biodynamica 7, 346–355.

Rey, L. R. 1957. Procedure for microscopic examination at low temperatures. Experientia 13, 201–202.

Rey, L. R. 1960. Study of the freezing and drying of tissues at very low temperatures. In Recent Research in Freezing and Drying. Edited by A. S. Parkes and A. V. Smith. Blackwell Scientific Publications, Oxford.

Rey L. R., and Bastien, M.-C. 1962. Importance of the preliminary freezing and sublimation periods. In Freeze-drying of Foods. Edited by F. R. Fisher. National Academy of Sciences, Washington.

RINFRET, A. P. 1962. Biochemical aspects of damage in freezing and freeze-drying of biological materials. *In* Freeze-drying of Foods. Edited by F. R. Fisher. National Academy of Sciences, Washington.

ROBSON, E. M., and ROWE, T. W. G. 1960. The physics of secondary drying. *In* Recent Research in Freezing and Drying. Edited by A. S. Parkes and A. V. Smith. Blackwell Scientific Publications, Oxford.

RUNDLE, R. E. 1955. The structure of ice. J. Phys. Chem. *59*, 680–682.

SCHAEFER, V. J. 1949. The formation of ice crystals in the laboratory and the atmosphere. Chem. Revs. *44*, 291–320.

SIMINOVITCH, D., and SCARTH, G. W. 1938. A study of the mechanism of frost injury to plant. Can. J. Research *C16*, 467–481.

SPONSLER, O. L., BATH, J. D., and ELLIS, J. W. 1940. Water bound to gelatin as shown by molecular structure studies. J. Phys. Chem. *44*, 996–1006.

STEPHENSON, J. L. 1960. Fundamental physical problems in the freezing and drying of biological materials. *In* Recent Research in Freezing and Drying. Edited by A. S. Parkes and A. V. Smith. Blackwell Scientific Publications, Oxford.

TAMMANN, G. 1925. The States of Aggregation. D. Van Nostrand, New York.

TAMMANN, G., and BÜCHNER, A. 1935. Supercooling of water and the linear crystallization velocity of ice in aqueous solutions. Z. anorg. u. allgem. Chem. *222*, 371–381.

THOMSON, J. 1861. On crystallization and liquefaction, as influenced by stresses tending to change of form in crystals. Proc. Roy. Soc. London *11*, 473–481.

TRESSLER, D. K., and EVERS, C. F. 1957. The Freezing Preservation of Foods. Vol. 1. Avi Publishing Co., Westport, Conn.

TURNBULL, D. 1950A. Formation of crystal nuclei in liquid metals. J. Appl. Phys. *21*, 1022–1028.

TURNBULL, D. 1950B. Kinetics of heterogeneous nucleation. J. Chem. Phys. *18*, 198–203.

TURNBULL, D., and FISHER, J. C. 1949. Rate of nucleation in condensed systems. J. Chem. Phys. *17*, 71–73.

TURNBULL, D., FISHER, J. C., and HOLLOMON, J. H. 1948. Nucleation. J. Appl. Physics *19*, 775–784.

URBIN, M. C., and WORLAND, M. C. 1961. Behavior of water phases in freeze drying. QM F & CI Contract No. A19-129-QM-1347, File No. A-332, Rept. No. 9.

WOOD, T. H., and ROSENBERG, A. M. 1957. Freezing in yeast cells. Biochim. Biophys. Acta. *25*, 78–87.

CHAPTER 15

Role and Behavior of Water in
Dehydration Processes

INTRODUCTION

Although the term dehydration could perhaps be correctly regarded as descriptive of the removal of water from a material by almost any means, including draining, pressing, centrifuging, freezing, surface activity, or extraction by a solvent, in the food industry the word is usually reserved for processes by which water is taken away in the vapor phase. It is in this sense that dehydration has undergone the most intensive theoretical and practical investigation, and the present discussion will be restricted to such processes. In accordance with the plan of the book, consideration will be given only to those reactions in which water is an essential participant. For details of specific production methods or other aspects of theory, the excellent survey of van Arsdel and Copley (1963, 1964) can be recommended.

The molecular steps involved in the dehydration of food materials can generally be stated as the change of water from a liquid, solid, or adsorbed state into the vapor phase and the subsequent transfer of the vapor out of the body of the product and into the atmosphere or a suitable collection trap. Generally, it is desirable that this be done with a minimum amount of change in the other characteristics of the fresh product, although some dried items (e.g., raisins, apricots, and certain kinds of dried fish) are given special organoleptic properties by sun-drying or other means of dehydration which are expected by the consumer and for which there is a definite demand. It is also important for economic reasons that the total energy put into the system be kept as low as possible consistent with maintenance of quality in the finished product.

In the drying of foodstuffs under substantially uniform conditions, the pattern of water loss can usually be divided into two stages, the constant-rate period in which the amount of water lost in each unit of time is the same, and the falling-rate period in which the amount of water lost decreases in each successive unit of time. The proportion of water present at the transition point between these two stages has been called the critical moisture content and given the symbol W_c.

Presumably, during the constant-rate period, the water in the food is subjected to no appreciable forces restraining it from vaporization other than the forces present in ordinary water substance or in an azeotrope. In the falling-rate step, some other restrictive forces must be operating. There may be a combination or sequence of these forces, and different forces may operate in different products or at different times in the same product. The elucidation of these forces is a matter of primary concern to investigators in the field of dehydration of foodstuffs.

Hygroscopic materials may enter the falling-rate stage almost at the start of dehydration while those largely insoluble substances containing water in voids may exhibit a constant-rate behavior pattern almost until they reach dryness, and will always do so until the layer next to the surface has reached substantially zero moisture content.

The physical factors immediately controlling the rate of water removal from a substance are the rate of heat transfer into the water, the rate of water transfer to the evaporating surface, and the partial pressure of water vapor in the layer of gas immediately surrounding the liquid, solid, or adsorbed water in the object.

HEAT TRANSFER

In most practical dehydration techniques, the limitation on speed of dehydration is the rate at which heat can be supplied to the point where the decrease in moisture content is occurring without intolerably raising the temperature of the body. There are three fundamental mechanisms of heat transfer: (1) conduction; (2) convection; and (3) radiation. Conduction is the transmittal of heat from one part to another part of the same body, or from one body to another in physical contact with it, there being no appreciable displacement of the particles of the bodies. Convection is the transfer of heat from one part to another within a gas or liquid, by the gross physical mixing of one part of the fluid with another. In unaided convection, the motion of the gas or liquid is entirely the result of the differences in density resulting from the differences in temperature. Forced convection is produced by mechanical operants such as fans. Radiation is the transfer of heat from one body to another which is not in contact with it by virtue of electromagnetic wave motion through space.

The fundamental differential equation for heat transfer by conduction is called Fourier's law:

$$q = \frac{dQ}{d\theta} = -kA\frac{dt}{dx}$$

where q is the rate of heat flow in B.t.u. per hour, Q is the quantity of heat transferred in B.t.u., θ is the time (hours), A is the area normal to the direction in which the heat flows (square feet), and dt/dx is the rate of change of temperature (°F.) with the distance (feet) in the direction of the flow of heat, that is, the temperature gradient. The factor k is called the thermal conductivity and is dependent upon the material through which the heat is flowing and upon temperature. The negative sign indicates that heat flows in the direction of lower temperatures.

The thermal conductivity of water at 32°F. is 0.343 and at 100°F. is 0.363 B.t.u./sq. ft/°F/ft.

The theoretical treatment of convection is difficult and complex. It will not be covered in this discussion. Standard texts on chemical engineering such as Chemical Engineers' Handbook (Perry *et al.* 1963) usually include more or less complete mathematical treatments of the transfer of heat by convection and should be consulted if further information on this point is desired.

The relative importance of the different mechanisms of heat transfer in any system varies with the temperature. At very low temperatures, heat transfer occurs chiefly by conduction. Each molecule transfers the excess kinetic energy which it possesses as a result of its higher temperature to an adjacent molecule of lesser energy so that successively colder layers are raised to higher levels. When the system contains fluid components, there operates the additional mechanism of convection, the transfer of energy through the movement of a relatively large amount (compared to the molecular scale) of gas or liquid. At elevated temperatures, the contribution of radiation may become appreciable. The molecules or atoms of one body undergo a sort of excitation caused by heat or some other method of energy input, and, in returning to a lower energy level, release the excess as radiant energy. This radiation may be captured by another body and transformed into heat.

The temperature at which radiation becomes an important contributor to heat transfer depends upon the emissivity of the surface. For large pipes losing heat by free convection, radiation contributes about half of the transferred energy at room temperature, while for thin wires having surfaces of low emissivity, radiation does not transfer half the energy until the wire reaches red heat (Gilmour *et al.* 1963). Emissivity is measured as the energy radiated from a unit area of a surface in unit time for unit difference in temperature between the emitting surface and surrounding bodies. For the centimeter-gram-second system, the dimensions of the emissive power are given in ergs per second per square centimeter with the radiating surface at 1° Absolute and the surroundings at Absolute zero. The normal total emissivity of a water surface is 0.95 at 32°F. and 0.963 at 212°F.

TRANSFER OF WATER

When any solid dries, two simultaneous processes must occur: (1) the transfer of heat to the evaporating liquid; and (2) the transfer of water as a liquid or vapor within the object and as a vapor from its surface. Factors determining the velocities of these processes govern the drying rate. Although heat transfer is probably the most frequent limiting factor, as previously stated, the rate of movement of water vapor can also influence the over-all velocity of drying in many cases.

In a discussion of the general phenomenon of evaporation, Langmuir and Schaeffer (1943) described it as a kinetic process involving the successive steps of (1) transport of water molecules to the surface, (2) passage across the surface region, and (3) diffusion through a quiescent layer of air overlying the liquid surface. Archer and LaMer (1954A and 1954B) considered each of the energy barriers to the evaporation steps proposed by Langmuir and Schaeffer in terms of a number of resistances. One conclusion reached as a result of this analysis was that a single resistance which is much greater than the others completely overshadows the others in determining the over-all rate of the evaporation process.

The driving force in evaporation is the difference in vapor pressure between the evaporating surface and the surrounding atmosphere. An empirically-derived "standard psychrometric formula" which relates vapor pressure to the temperature drop at an evaporating surface is:

$$P_{sw} - P_w = \frac{P}{2,730} (t_a - t_w) \left(1 + \frac{t_w - 32}{1,571}\right)$$

where t_a is the air temperature in °F., t_w is the wet-bulb temperature in °F., P is the barometric pressure in atmospheres, P_{sw} is the vapor pressure of water at the wet-bulb temperature in atmospheres, and P_w is the partial pressure of water vapor in the air in atmospheres. A simplified formula accurate enough for most purposes has been quoted by Krischer (1939):

$$P_{sw} - P_w = \frac{P(t_a - t_w)}{2,720}$$

Two different zones can exist in a drying object: (1) the saturated region, in which the water molecules form a three-dimensional continuum of some form having the essential properties of liquid water or ice; and (2) the unsaturated zone consisting in effect of single water molecules bound to a non-aqueous component of the object by forces differing in type and energy content from the intermolecular forces in pure water substance. At some time during the period of drying to completion, the saturated zone will be completely replaced by the second type of zone. Any non-homogeneous object may consist of many zones of each type.

Within the saturated zone water may be transferred by: (1) capillary flow; (2) diffusion of liquid water; (3) vaporization with or without subsequent condensation; (4) flow caused by shrinkage and pressure gradients; and (5) gravitational flow. In the unsaturated zone, transfer of water must occur by displacement of single molecules which may be followed by adsorption or condensation within the drying substance or by loss to the surrounding atmosphere.

Internal moisture transfer by diffusion in drying solids has been studied extensively. Sherwood (1929) solved the diffusion equation for the falling-rate period in a slab assuming that the surface is dry or at its equilibrium moisture content and that the initial moisture distribution is uniform. Evaporation was assumed to occur from both faces of the slab. For these conditions, the following equation was obtained:

$$\frac{W - W_e}{W_c - W_e} = \frac{8}{\pi^2}\left[e^{-D_l\theta(\pi/2d)^2} + \frac{1}{9}e^{-9D_l\theta(\pi/2d)^2} + \frac{1}{25}e^{-25D_l\theta(\pi/2d)^2} + \cdots\cdots\right]$$

where

W = average moisture content (dry basis) at any time θ

W_c = average moisture content (dry basis) at the start of the falling-rate period.

W_e = average moisture content (dry basis) in equilibrium with the environment

D_l = liquid diffusivity, in square feet per hour

θ = time from the start of the falling-rate period, hour

d = one-half the thickness of the solid layer through which diffusion occurs

For a long drying time, the above equation can be simplified to the limiting form:

$$\frac{W - W_e}{W_c - W_e} = \frac{8}{\pi^2}e^{-D_l\theta(\pi/2d)^2}$$

and then differentiated to give the drying rate $dW/d\theta$:

$$\frac{dW}{d\theta} = -\frac{\pi^2 D_l}{4d^2}(W - W_e)$$

From this equation it can be seen that, under the postulated conditions the drying rate is: (1) inversely related to the square of the thickness of the slab; (2) directly related to the diffusivity; and (3) directly related to the free water content as expressed by the term $(W - W_e)$.

Sherwood's equation assumed a slab where thickness is small compared to the other two dimensions. The mathematical analysis of other shapes has been discussed by Crank (1956).

Diffusion equations were applied with some success to the drying of clay and lumber. According to Ceaglske and Hougen (1937) the correspondence between theory and practice in these cases arose from a fortuitous counterbalancing of errors. They characterized as incorrect the assumption that the rate of drying of a granular solid is controlled by diffusion, giving as their reason the observation that water will pass from a region of lower concentration in coarse sand to one of higher concentration in fine sand. Instead of diffusion, the controlling factor is a complex suction-water concentration relationship. In the drying of a body of sand filled with water, there is an initial constant rate period of moisture removal until suction at the surface exceeds 8.6 cm. Then the rate of drying falls off linearly with the moisture content of the sand because the actual wetted area at the surface diminishes. The slope of the curve increases with increasing thickness of the drier layer. When the water film is broken, the surface becomes dry and the dry area extends progressively below the surface. Drying then proceeds by transfer of vapor through this dry area. Therefore, to retard drying of sand, sprinkle coarse sand on the surface, to hasten drying, sprinkle fine sand on the surface. Ceaglske and Hougen described methods for calculating the moisture distribution and average moisture content for any thickness of a granular solid and for calculating the drying rate curves during the two falling periods from the value during the constant-rate period.

These and other studies have shown that the movement of liquid water within a porous drying object can occur as a result of capillary action. Calculations in capillary theory have been made using the assumption that the voids are spaces between spheres of uniform radius. As water evaporates from the voids, the curvature of the liquid surfaces changes, creating a so-called suction potential which is an index of the force tending to draw water to the surface from the interior. The suction potential, P_s (in centimeters of water) can be calculated from the following equation (Bagnoli et al. 1963):

$$P_s = \frac{X\sigma}{r\rho g}$$

where σ is the surface tension in dynes per centimeter, ρ is the density of water in grams per cubic centimeter, g is the acceleration of gravity 980 cm./sec./sec., r is the sphere radius in centimeters, and X is a packing factor equal to 12.9 for rhombohedral and 4.8 for cubical packing.

As the moisture evaporates, the surface menisci recede until the suction potential reaches a value given by the above equation. At that time, the pores at the surface will open, air will be drawn in, and the moisture will redistribute with a slight lowering of the suction potential. The cycle is then repeated.

A modified form of the Poiseuille equation for laminar flow can be used to determine the pressure difference required to overcome friction in a bed of uniform, spherical, non-porous particles. The equation is:

$$H_p = \frac{K_2 \mu V_f \rho_t^2 S^2 (1 - F)^2 d}{g \rho F^3}$$

where:

H_p = pressure head, centimeter
K_2 = Kozený constant (five for spherical particles)
ρ = density of water, grams per cubic centimeter
V_f = flow velocity, centimeter per second
μ = viscosity of water, poise
S = specific surface of the particles, square centimeter
ρ_t = density of particles, in gram per cubic centimeter
F = porosity, dimensionless
d = depth of bed, centimeter
g = acceleration of gravity, 980 cm./sec./sec.

If there is rhombohedral packing of the spheres, the substitution of 3.15 for $(1-F)^2/F^3$ and $9/r^2$ for $\rho_t^2 S^2$ can be made, giving the equation $H_p = 141.8 \, (\mu V_f d/g \rho r^2)$. Similar calculations can be made for other types of packing by using appropriate values for F, ρ_t, and S. To maintain flow to the surface of the drying solid, $P_1 = P_2 + d_2 + H_p$ where P_1 is the suction potential of the surface, P_2 the suction potential at a depth of d_2 centimeter, and H_p is the loss of head or frictional resistance of the bed over a distance d. Substituting $P_1 - P_2 - d_2$ for H_p in the foregoing equation, we obtain:

$$\frac{P_1 - P_2 - d_2}{d} = 141.7 \, \frac{\mu V_f}{g \rho r^2}$$

but $V_f = (1/y)(dW/d\theta)$, where y is the fraction of pores at the surface, so that

$$\frac{dW}{d\theta} = \frac{y r^2}{3.25 \times 10^{-5} d} (P_1 - P_2 - d_2)$$

and the drying rate may be calculated if P_1, P_2, and y are known.

Corben and Newitt (1955) and King and Newitt (1955) showed that the drying characteristics of moist porous granular materials are consistent with the capillary theory of drying. The form of the drying rate curves differs from those for non-porous granular materials mainly through the capillary action of the internal pores. The rate of drying during the constant rate period is higher for porous than for non-porous material, and, after the critical moisture content is reached, there follows an initial period having a very rapid reduction in drying rate and a prolonged second period in which the drying rate falls slowly.

SPECIAL FORMS OF DRYING

Freeze-drying

In freeze-drying (also called lyophilization and sublimation drying) some of the harmful effects of freezing are superimposed upon those of drying. However, advantages resulting from the use of low temperatures during the removal of water, such as inhibition of solute diffusion and concentration and the retention of shape and size due both to the hardening of the structural components of the raw material and the prevention of surface effects which occur when liquid water is present, often outweigh the disadvantages. The factors affecting freezing and the effects of this process upon foodstuffs described in another chapter.

To secure the full advantage of freeze-drying, it is desirable that all eutectics be frozen before drying is begun. Impractically low temperatures may be necessary to achieve this status. Greaves (1954) found that blood serum has a eutectic below $-76°F$. The vapor pressure of the aqueous phase at this temperature is so low that drying would take a very long time to complete; therefore serum is dehydrated at a higher temperature. Appreciable damage is avoided because the quantity of eutectic and the amount of protein in contact with it is small.

As in any form of drying, it is necessary to maintain a lower water activity around the ice and frozen eutectics than exists in the un-frozen aqueous systems. The activity of water in equilibrium with pure ice at various temperatures, T, below $32°F$. can be calculated from an equation given by Lewis and Randal (1923): $\log a = -0.004211\ T - 0.0000022\ T^2$, where a is the water activity or relative vapor pressure. Water in equilibrium with pure ice at a temperature of $-4°F$. will exhibit an activity of 0.822.

Low water activity in the drying system can be maintained by including in it a trap or condenser held at a very low temperature so that the ice which is collected will have a lower vapor pressure than the water in the food, by continuously pumping the evolved vapor out of the drying chamber, or by continuously passing the combined vapors from the chamber through an efficient desiccant. The latter method is not used commercially. The most common procedure is the use of an evacuated chamber together with a condenser placed in the near vicinity of the drying product. The close proximity of the condenser to the food shortens the path of the water molecules coming from the drying material and contributes to the efficiency and speed of the dehydration process.

It is necessary to supply heat to the water molecules to overcome the intermolecular binding forces, i.e., to supply the latent heat of sublimation. In most cases, it is the rate of heat transfer through the surface-dry layers into the mass of frozen aqueous phase which is the principal rate-limiting factor in freeze dehydration. If external heating is not used, the temperature of the drying material will constantly decrease, with related decreases in drying rate and in vapor pressure over the ice. Heat can be supplied by placing the objects on a heated platen, or by radiation. It is highly desirable that the heat input be adjusted so that the aqueous phase in the food never becomes liquid. Rather high platen temperatures can sometimes be used in the earlier stages of drying since the rapid sublimation consumes energy at a rate sufficiently great to offset the high input resulting from the steep temperature gradient between the platen surface and the frozen core of the product. As the moisture content becomes lower, the water molecules are lost at a lower rate, and a platen temperature which was formerly satisfactory may result in an excess supply of energy with consequent melting of the eutectics and ice. Of course, as the product dries, the water level recedes from the platen and the greater thickness of dry material retards the rate of heat transfer. Under certain conditions it has been shown to be possible to burn the surface of a partially dried product without melting the ice, so effective are the insulating properties of the dried layer.

After the ice has been removed the porous piece will still retain adsorbed water [usually about seven per cent (Robson and Rowe 1960)], removal of which will be very slow at the temperatures of freeze drying. It is, however, necessary to remove most of the adsorbed water in order to obtain adequate storage stability. An increase in the temperature of the object is permissible after the ice has been removed because collapse due to the surface effects of liquid water is no longer a problem. Bin-drying or in-package desiccation can also be used to remove most of the remaining moisture.

In some experiments on freeze drying of carrot, apple, and lean beef, it was found that the rate of drying was correlated with (1) the pressure in the chamber, (2) the difference between the temperatures of the platen and the food, and (3) the difference between the vapor pressure of ice at the temperature of the food and its vapor pressure at the condenser temperature. The principal determinants of the drying rate, in their experiments, were the temperature of the platen and the size of the piece.

Considering the simplified case where the food materials are in the form of spheres and remaining moisture content at any time is contained entirely in a frozen core of spherical shape, other workers have been able to develop a mathematical treatment of the drying rate. In this simplified situation, the resistance to diffusion of the water vapor from the constantly shrinking surface of the core through the dried porous shell is the only significant resistance to the moisture diffusion, and the ratio of the difference between the vapor pressure of the product ice and that of the condenser ice to the drying rate should be a linear function of the expression

$$\frac{1 - (1 - x)^{1/3}}{(1 - x)^{2/3}}$$

where x is the fraction of the moisture content which has been removed. The correlation can be formulated as follows:

$$\frac{P_{ip} - P_{ic}}{dx/d\theta} = \frac{k_2 r_0}{3} + \frac{k_1 r_0^2 [1 - (1 - x)^{1/3}]}{3 (1 - x)^{2/3}}$$

where

$\quad P_{ip} =$ vapor pressure of ice at the temperature of the core
$\quad P_{ic} =$ vapor pressure of ice at the temperature of the condenser
$\quad r_0 =$ initial radius of the object in centimeter
k_1 and $k_2 =$ constants having the dimensions (millimeter Hg) (hour)/ square centimeter

Data from the experiment on carrot dice indicated that the value of the term $k_2 r_0/3$ was near 1.0 for the conditions used.

Drum-drying

This process relies upon the evaporation of water from a thin film of the raw material spread on a hot drum. Decline in moisture content is quite rapid due to the large amount of surface exposed and the high temperature of the drum, both leading to rapid vaporization of the water and diffusion of the vapor. However, a considerable amount of heat damage can occur. In milk, the product which is most extensively processed by both spray- and drum-drying techniques, good quality spray-dehydrated product is always superior to the best drum-dried milk in terms of lower protein denaturation and other indices of heat damage. The process is also obviously restricted to those substances which can be reduced to a homogeneous liquid. There has not been published any thorough discussion of the principles of drum-drying, although several worthwhile articles on specific applications of the process are available (e.g., Cording *et al.* 1957).

Foam-mat Drying and Puff-drying

Puffing is a phenomenon observed in the later stages of the vacuum drying of initially rather dilute substances such as fruit juices, plasma, etc.

Evidently, the progessive decrease in moisture content ultimately leads to a viscosity of the medium sufficiently high to maintain stable bubbles. A recent discussion has been published by Fixari *et al.* (1959).

Foam-mat drying includes a whipping or aeration step to develop a mass of bubbles which is then subjected to dehydrating conditions (see, for example, Lawler 1962). Vacuum conditions are generally not used in foam-mat drying and the temperatures involved are relatively low. The advantages of these conditions are that the liquid-gas interface is enormously increased and fragmentation or powdering of the dried material is facilitated. On the other hand, heat transfer is impeded by the large volume of gas present in the mass. The finished product is very porous and absorbs water without the formation of agglomerates, being superior in these respects to drum-dried flakes or spray-dried powders. Case-hardening and similar phenomena are completely eliminated.

In-package Desiccation

This process involves inclosing the partially dehydrated material in a container with a relatively small amount of chemical desiccant. For foods, the desiccant is frequently either silica gel or calcium oxide since these substances are among the most innocuous of the available drying agents so that accidental ingestion of the package or part of its contents will be less dangerous. In-package desiccation is very gentle so far as temperature effects are concerned, but the slowness of the interchange of moisture may permit spoilage reactions to occur before the water activity is reduced to a safe level. See Hendel *et al.* (1958) for a complete discussion of this method of drying.

Spray-drying

The theoretical analysis of heat transfer and moisture diffusion under the conditions existing during spray-drying has proved to be very difficult. The material undergoes strong shearing forces as it passes through the atomizer with the result that intense turbulence is set up in the liquid and these currents cannot be expected to be consistent from droplet to droplet. The droplets pass through regions of constantly changing water-vapor pressure and temperature, making any kind of static approximations inexact. It has been shown experimentally that the size, shape, and distribution in space of the droplets as established by the atomizer characteristics have a major influence on the speed of drying and the quality of the finished product, evidently through the effect these factors have on the rate of water diffusion from the surface of the droplet and from the interior of the droplet to its surface. The statistical distribution of droplet sizes produced by various kinds of atomizers and the effects of viscosity, density, and surface tension of the liquid have been thoroughly examined (Marshall 1954). There are many publications dealing with spray-drying of specific products, e.g., Coulter (1955).

The droplets evaporate with a surface temperature corresponding to the saturated solution even though the droplet concentration, as a whole, is less than saturation (Ranz and Marshall 1952). According to Duffie and Marshall (1953), virtually every material that has been spray-dried consists of hollow-spherical-shaped particles. This shape results from: (1) forming of films causing puffing or ballooning; (2) rate of evaporation exceeding the diffusion rate of salts back into the particle creating internal voids after dryness is reached; (3) capillary action of the material on the drop surface withdrawing liquids and solids to the surface creating subatmospheric pressures within the particle, and (4) presence of entrained gas in the liquid feed. The bulk density of spray-dried materials studied by these workers (e.g., organic dyestuffs, sodium silicate, and sodium chloride) decreased with increases in drying temperature, probably attributable to an increasing tendency of the particles to expand on drying and to earlier formation of vapor-impervious films on the drop surface. Increases in feed temperature caused either a small increase or a small decrease in the bulk density. Increased feed temperature caused a slightly increased drying rate and, by affecting the viscosity, influenced the particle size resulting from atomization. The effect of solids on bulk density depended largely on whether the increased feed concentration affected the tendency of the particles to expand.

Duffie and Marshall classified materials into three categories based on observation of dried particle characteristics: (1) film-forming materials such as sodium silicate; (2) crystalline materials such as most inorganic salts; and (3) intermediate materials having limited film-forming tendencies, such as water-dispersable organic dyes.

Materials spray-dried at low temperatures (not over 86°F.), as in the method described by Ziemba (1962), are in the usual form of spheres but these are not hollow as are the particles spray-dried by other methods. This observation tends to confirm the theory that hollow droplets are formed when the initial drying rate is so high that a very steep gradient of moisture content is set up in a droplet to offset the high rate of heat transfer.

BIBLIOGRAPHY

ANON. 1961. Guide and Data Book No. 1. American Society of Heating, Refrigerating, and Air-Conditioning Engineers, New York.

ARCHER R. J., and LaMER, V. K. 1954A. The effect of monolayers on the evaporation of water. Ann. N. Y. Acad. Sci. 58, 807–829.

ARCHER, R. J., and LaMER, V. K. 1954B. The rate of evaporation of water through fatty acid monolayers. J. Phys. Chem. 59, 200–208.

BAGNOLI, E., FULLER, F. H., and NORRIS, R. W. 1963. Humidification and drying. In Chemical Engineers' Handbook. Fourth Edition. Edited by R. H. Perry, C. H. Chilton, and S. D. Kirkpatrick. McGraw-Hill Book Co. New York.

BATES, H. T. 1960. Economical drying of solids. Chem. Eng. Progr. 56, No. 7, 52–57.

BECKETT, L. G. 1954. The effects of residual moisture in frozen-dried materials, and its management. In Biological Applications of Freezing and Drying. Edited by R. J. C. Harris. Academic Press, New York.

BOSWORTH, R. C. L. 1956. Transport Processes in Applied Chemistry. John Wiley and Sons, New York.

BUCHANAN, T. J., HAGGIS, G. H., HASTED, J. B., and ROBINSON, B. G. 1952. The dielectric estimation of protein hydration. Proc. Royal Soc. London 213A, 379–381.

BULL, H. B. 1944. Adsorption of water vapor by proteins. J. Am. Chem. Soc. 66, 1499–1507.

CARMAN, P. C. 1948. Molecular distillation and sublimation. Trans. Faraday Soc. 44, 529–536.

CARMAN, P. C. 1956. Flow of Gases Through Porous Media. Academic Press, New York.

CARSLAW, H. S., and JAEGER, J. C. 1947. Conduction of Heat in Solids. Clarendon Press, Oxford.

CARTER, J. W. 1960. Adsorption drying of gases. Brit. Chem. Eng. 5, 472–476, 552–555, 625–631.

CEAGLSKE, H. H., and HOUGEN, O. A. 1937. Drying granular solids. Ind. Eng. Chem. 29, 805–813.

COMINGS, W. E., and SHERWOOD, T. K. 1934. The drying of solids. VII. Moisture movement by capillarity in drying granular materials. Ind. Eng. Chem. 26, 1096–1098.

CORBEN, R. W., and NEWITT, D. M. 1955. Mechanism of drying of solids. VI. The drying characteristics of porous granular material. Trans. Inst. Chem. Engrs. 33, No. 1, 52–63.

CORDING, J. JR., WILLARD, M. J., JR., ESKEW, R. K., and SULLIVAN, J. F. 1957. Advances in the dehydration of mashed potatoes. Food Technol. 11, 236–240.

COULTER, S. T. 1955. Evaporation of water from milk by spray drying. J. Dairy Sci. 38, 1180–1184.

COULTER, S. T., JENNESS, R., and GEDDES, W. F. 1951. Physical and chemical aspects of the production, storage, and utility of dry milk products. Advances in Food Research 3, 45–118.

CRANK, J. 1956. The Mathematics of Diffusion. Clarendon Press, Oxford.

DELANEY, L. J., HOUSTON, R. W., and EAGLETON, L. C. 1964. The rate of vaporization of water and ice. Chem. Eng. Sci. 19, 105–114.

DOLE, M., and McLAREN, A. D. 1947. The free energy, heat, and entropy of sorption of water vapor by proteins and high polymers. J. Am. Chem. Soc. 69, 651–657.

DUCKWORTH, R. B., and SMITH, G. M. 1961. The diffusion of glucose during vegetable dehydration. J. Sci. Food Agr. 12, 490–492.

Duffie, J. A., and Marshall, W. R., Jr. 1953. Factors influencing the properties of spray-dried materials. II. Chem. Eng. Progr. *49*, 480–490.

Fish, B. P. 1957. Diffusion and Equilibrium Properties of Water in Starch. H. M. Stationery Office, London.

Fish, B. P. 1958. Diffusion and thermodynamics of water in potato starch gel. *In* Fundamental Aspects Dehydration Foodstuffs, Papers Conf. Aberdeen *1958*, 143–147.

Fisher, E. A. 1935. Some fundamental principles of drying. J. Soc. Chem. Ind. *54*, 343–348.

Fixari, F., Conley, W., and Bard, G. 1959. Continuous high vacuum drying techniques. Food Technol. *13*, 217–220.

Gane, R. 1941. The water content of dried foodstuffs as a function of humidity and temperature. Low Temp. Research Sta. Interim Record Memo. No. 111.

Gersh, I., and Stephenson, J. L. 1954. Freezing and drying of tissues for morphological and histochemical studies. *In* Biochemical Applications of Freezing and Drying. Edited by R. J. C. Harris. Academic Press, New York.

Gilmour, C. H., Hottel, H. C., and Weger, E. 1963. Heat transmission. *In* Chemical Engineers' Handbook. Fourth Edition. Edited by R. H. Perry, C. H. Chilton, and S. D. Kirkpatrick. McGraw-Hill Book Co., New York.

Goff, J. A., and Gratch, S. 1946. Low-pressure properties of water in the range −160° to 212°F. J. Am. Soc. Heating Ventilating Eng. *52*, 95–121.

Goldblith, S. A., Karel, M., and Lusk, G. 1963A. The role of food science and technology in the freeze dehydration of foods. I. Food Technol. *17*, 139–144.

Golblith, S. A., Karel, M., and Lusk, G. 1963B. Role of food science and technology in the freeze dehydration of food. II. Food Technol. *17*, 258–263.

Goodman, W. 1938. New tables of the psychrometric properties of air-vapor mixtures. Heating Piping Air-conditioning *10*, No. 1, 1–4, 119–122.

Gortner, R. A. 1949. Outlines of Biochemistry. Third Edition. John Wiley and Sons, New York.

Greaves, R. I. N. 1954. Theoretical aspects of drying by sublimation. *In* Biological Applications of Freezing and Drying. Edited by R. J. C. Harris. Academic Press, New York.

Greaves, R. I. N. 1962. Minimizing product damage when drying biological solutions. *In* Freeze-drying of Foods. Edited by F. R. Fisher. National Academy of Sciences, Washington.

Gregory, S. A. 1960. Recent developments in adsorption drying. Ind. Chemist *36*, 479–484.

Grubenmann, M. 1958. I-x Diagrams for Humid Air. Springer-Verlag, Berlin.

Hamm, R. 1960. Biochemistry of meat hydration. Advances in Food Research *10*, 355–360.

Hendel, C. E., Legault, R. R., Talburt, W. F., Burr, H. K., and Wilke, C. R. 1958. Water vapor transfer in the in-package desiccation of dehydrated foods. *In* Fundamental Aspects of the Dehydration of Foodstuffs.

Hickman, K. C. 1954. Maximum evaporation coefficient of water. Ind. Eng. Chem. *46*, 1442–1446.

KAWASAKI, K. 1962. On the variation of wettability of organic solids in contact with water. J. Colloid Sci. *17*, 169–177.

KING, A. R., and NEWITT, D. M. 1955. Mechanism of drying of solids. VII. Drying with heat transfer by conduction. Trans. Inst. Chem. Engrs. *33*, 64–69.

KRISCHER, O. 1939. Physical problems in the drying of solid porous materials. Chem. App. *26*, 17–23.

KUPRIANOFF, J. 1962. Some factors influencing the reversibility of freeze-drying of foodstuffs. *In* Freeze-drying of Foods. Edited by F. R. Fisher. National Academy of Sciences, Washington.

LANGMUIR, I. and SCHAEFFER, V. J. 1943. Rates of evaporation of water through compressed monolayers on water. J. Franklin Inst. *235*, 119–162.

LAWLER, F. K. 1962. Foam-mat drying goes to work. Food Eng. *34*, No. 2, 68–69.

LAZAR, M. E., BARTA, E. J., and SMITH, G. S. 1963. Dry-blanch-dry method for drying fruit. Food Technol. *17*, 1200–1202.

LAZAR, M. E., and POWERS, M. J. 1959. Method for preparing unsulfured dehydrated fruits. U. S. Pat. No. 2,895,836. July 21.

LEWIS, G. N., and RANDALL, M. 1923. Thermodynamics. McGraw-Hill Book Co., New York.

LIVINGSTON, H. K. 1944. Cross-sectional areas of molecules adsorbed on solid surfaces. J. Am. Chem. Soc. *66*, 569–573.

MARSHALL, W. R., JR. 1953. Drying. Ind. Eng. Chem. *45*, 47–54.

MARSHALL, W. R., JR. 1954. Atomization and Spray Drying. Am. Inst. Chem. Eng. Chem. Engineering Progr. Monograph Series No. 2.

MARSHALL, W. R., JR., and FRIEDMAN, S. J. 1950. Drying. *In* Chemical Engineers' Handbook. 3rd Edition. Edited by J. H. Perry. McGraw-Hill Book Co., New York.

MARSHALL, W. R., JR., and HOUGEN, O. A. 1942. Drying of solids by through-circulation. Trans. Am. Inst. Chem. Eng. *38*, 91–121.

MARVIN, C. F. 1941. Psychrometric tables for vapor pressure, relative humidity, and temperature of the dew-point. U. S. Weather Bureau Publ. *235*.

McADAMS, W. H., HOTTEL, H. C., COLBURN, A. P., and BERGELIN, O. P. 1950. Heat transmission. *In* Chemical Engineers' Handbook. 3rd Edition. Edited by J. H. Perry. McGraw-Hill Book Co., New York.

McCORMICK, P. Y. 1961. Drying. Ind. Eng. Chem. *53*, 583–590.

MERYMAN, H. T. 1962. Introductory survey of biophysical and biochemical aspects of freeze-drying. *In* Freeze-drying of Foods. Edited by F. R. Fisher. National Academy of Sciences, Washington.

NEMETHY, G., and SCHERAGA, H. A. 1962. Structure of water and hydrophilic bonding in proteins. I. A model for the thermodynamic properties of liquid water. J. Chem. Phys. *36*, 3382–3400.

NOVAK, A. F., and RAMACHANDRA RAO, M. R., 1964. Carbon monoxide inactivates enzymes. Food Processing *25*, No. 9, 92–95, 98.

ONSAGER, L. 1945. Theories and problems of liquid diffusion. Ann. N. Y. Acad. Sci. *46*, 241–265.

RANZ, W. E., and MARSHALL, W. R., JR. 1952. Evaporation from drops. Chem. Eng. Progr. *48*, 141–146, 173–180.

ROBSON, E. M., and ROWE, T. W. G. 1960. The physics of secondary drying. *In* Recent Research in Freezing and Drying. Edited by A. S. Parkes and A. U. Smith. Blackwell Scientific Publications, Oxford.

SARAVACOS, G. D., and CHARM, S. E. 1962. A study of the mechanism of fruit and vegetable dehydration. Food Technol. *16*, 78–81.

SHERWOOD, T. K. 1929. The drying of solids. Ind. Eng. Chem. *21*, 12–116, 976–980.

SPONSLER, O. L., BATH, J. D., and ELLIS, J. W. 1940. Water bound to gelatin as shown by molecular structure studies. J. Phys. Chem. *44*, 996–1006.

URBIN, M. C., and WORLAND, M. C. 1961. Behavior of water phases in freeze drying. QMF & CI Contract No. A19-129-QM-1347, file No. A-332, Rept. No. 9.

VAN ARSDEL, W. B. 1963. Food Dehydration. I. Principles. Avi Publishing Co., Westport, Conn.

VAN ARSDEL, W. B., and COPLEY, M. J. 1964. Food dehydration. II. Processes and Products. Avi Publishing Co., Westport, Conn.

WIERBICKI, E., and DEATHERAGE, F. E. 1958. Determination of water-holding capacity of fresh meats. J. Agr. Food Chem. *6*, 387–392.

ZIEMBA, J. V. 1962. Now—drying without heat. Food Eng. *34*, No. 7, 84–85.

CHAPTER 16

The Effect of Water on Food Texture

INTRODUCTION

It is readily apparent that variations in moisture content have a pronounced influence on the texture of most foodstuffs. In many cases this effect is quite critical, changes of a per cent or so within a certain range making the difference between an acceptable and an unsatisfactory product. The effect of the moisture content on texture of foods is perceived in three principal ways: (1) the free aqueous phase which is pressed out when the product is chewed—variations in amount and viscosity of the fluid (which are, of course, related to the moisture content) leading to impressions of varying degrees of juiciness; (2) surface stickiness—most obvious in hard candies and similar foods, but actually a factor in many other products; and (3) the results of interactions with structural components—the proteins and carbohydrates forming the more-or-less rigid structure of foodstuffs vary widely in elasticity and other physical properties depending on the extent to which they are hydrated. In this chapter the subject will be explored by discussing the effect of moisture variations on the texture of several representative food products.

FRESH FRUITS AND VEGETABLES

The crispness of lettuce, watermelon, celery, and many other fresh fruits and vegetables is principally a function of the turgor of the cell. A "crunchiness" due to the fiber content may remain after all turgor has been abolished (as by wilting), but true crispness depends upon the presence of many distended, liquid-filled cells. In many of these foods, the cell wall itself contains a relatively insignificant amount of rigid structural elements and contributes in only a minor way to texture. Thus, for all practical purposes form and texture will depend upon the presence of rather dilute solutions enclosed in selectively permeable membranes. This effect can be seen best in some of the melons, where loss of selective permeability, through heat damage or other means, results in complete collapse of the tissue and disappearance of the characteristic texture. On the other hand, celery and the rind of watermelon, for example, lose crispness under these circumstances but retain a tough texture and much of the original volume. The cell walls in such foods contain enough rigid structural elements to maintain cell form in the absence of turgor.

Water can be lost through the semipermeable membrane in the vapor phase with a consequent reduction in turgidity. The rate at which evaporation occurs is related to the water activity of the contacting atmosphere. Numerous papers have contained information on the greater wilting that occurs in drier atmospheres and the apparent preservative effect on texture that is observed when crisp products such as lettuce are kept in water or in storage areas where the humidity is maintained at high levels. Somewhat equivalent results can be obtained by wrapping the product in films resistant to moisture vapor transfer, in which case the product adjusts its own atmosphere. This procedure is now being used extensively with lettuce and other fresh vegetables.

In a study of possible methods for extending the storage life of fresh vegetables, Cook *et al.* (1958) found that cabbage retained quality better at 32°F. than at 38°F., remaining "fresh and turgid" for eight weeks at the lower temperature. Celery also retained crispness longer at 32°F. A similar response of lettuce and tomatoes to storage conditions was observed. Packing the foods in polyethylene bags decreased wilting. A similar investigation, reported by Parsons (1959), showed that cabbage stored equally well at 32°F. and 38°F. but deteriorated rapidly at 45°F. When the relative humidity was 92%, wilting occurred unless the cabbages were stored in crates with polyethylene linings. Wilting of lettuce at low temperatures, observed by Parsons and Wright (1956) was attributed by them to the difficulty of maintaining a sufficient amount of water vapor in the surrounding air at these temperatures. When the product is protected from air of low relative humidity by polyethylene wrapping, low temperatures assist in maintaining crispness.

There are no publications in which the turgor pressure or hydrostatic pressure is quantitatively related to measurements of textural characteristics. Weckel *et al.* (1959) measured "turgidity or firmness" by forcing two parallel wires 0.019 in. in diameter through a one millimeter midsection of canned potato, but it is clear that their use of the word is not in accord with the definition of turgidity used here. Turgor pressure is frequently determined by immersing cells in liquids of varying osmotic pressure for a brief time and then observing the cells microsopically for signs of plasmolysis. The turgor pressure is considered to be approximately equivalent to the osmotic pressure which plasmolyses 50% of the cells. Such measurements might be correlated to perceived texture. It might also be possible to secure meaningful data relating turgor pressure to textural characteristics by rapidly bringing pieces of the material to about 140° to 150°F., thus destroying the selective permeability of the cell membrane, and then comparing the resistance to crushing or penetration with that of other cells not so treated. Of course, other changes occur in this temperature range, and measurements would have to be made rapidly to minimize the effects of enzymatic changes. In spite of the complicating factors, data secured by the suggested procedure should allow a worthwhile estimate to be made of the contribution of turgor pressure to perceived texture. Allowances for the decrease in volume due to loss of turgor pressure would have to be made.

Nuts comprise a group of crisp foods about whose texture very little of a scientific nature has been written. It is probable that crispness in nuts is due to cellular distension by the large amounts of oil which are present. Pecans, for example, contain less than five per cent water and more than 70% oil. The oil is contained in pockets or vacuoles and it is quite likely that the small volume of cellular material surrounding these masses is hydrated to about the same extent as in other seed structures. Heat treatment seems to have a stiffening effect—the crispness is increased—in nuts, contrary to what is seen in most other turgid products.

In sweet corn and in some legumes, the moisture content is a reliable indicator of texture since the per cent dry matter tends to increase as the product becomes tougher and more mature. According to Huelsen (1954), the general practice of sweet corn canners is to harvest 71 to 74% moisture corn for the whole grain pack and sort out the drier more mature ears for the cream style pack, where the texture is less critical. Loss of moisture with increasing maturity is accompanied by "denting" of the kernels in sweet corn, and this feature can also be used as an indication of the textural quality of the grain.

SAUSAGE

Sausages are of two general kinds, the high-moisture types which are made from a so-called emulsion composed of added water and the meat constituents (sometimes with added fat), and the low-moisture types which do not contain added water and may be partially dried by smoking or like processes. The texture of both kinds of sausages is affected by the moisture content, although the particle size, the inherent fiber toughness, the fat content, and processing conditions all play some part in determining the consistency of the finished product. In the high-moisture types with small particle size, the water content is doubtless of greater influence on the over-all texture than in dry products where the meat pieces are large enough to be perceived separately.

Regulations of the Meat Inspection Division of the U. S. Department of Agriculture specify that the moisture in the final cooked sausage must not exceed four times the meat protein (by analysis) plus ten per cent, i.e., moisture $< 4P + 10$. Since meat will normally have a moisture to protein ratio of about 4 to 1, about ten per cent additional moisture can be added in most cases. However, in sausage that is not heat processed the limit is lower, four times the protein plus three per cent. Swift *et al.* (1954) indicate that the tenderness and juiciness of frankfurters and bologna are directly related to the content of both fat and moisture. Equivalent levels of juiciness and tenderness can be attained by substituting one for the other, but these textural qualities are more readily altered by adjusting the moisture content. There is ample evidence that in regard to consumer preferences there is an optimum tenderness and juiciness for frankfurters (Wilson 1960). The Committee on Sausage of the American Meat Institute has studied limitations on additions of moisture and fat for the production of acceptable products. Although moisture in excess of the limited amount could be added, the acceptable amount was less in about half of the cases and appeared to be related to the protein content.

The water-binding capacity of meat can be increased and the texture altered by additives such as tetrasodium pyrophosphate, sodium acid pyrophosphate, hexametaphosphate, and sodium polyphosphate. Even sodium chloride can, under certain conditions, increase imbibition capacity. The phosphates also play a part in fat emulsification. The "emulsion" of sausage consists of a matrix of protein and water encapsulating particles of fat. Fluid retention at 32°F. from sodium chloride and other neutral salt solutions depends on the degree of ion absorption, anions being retained preferentially. The amount of fluid retained corresponds well with anion absorption.

Grau and Hamm (1953) described a method for determining the water holding capacity of meat. It involved pressing a small piece of muscle tissue between filter paper and measuring the weight lost by the meat. By using this method, Wismer-Pedersen (1959) showed that much variation occurred between different hogs in imbibition capacity of muscles.

EMULSIONS

The proportion of oil to water in oil-in-water emulsions tends to determine their rigidity. As increasing amounts of oil are added, the film of the aqueous phase becomes thinner. At some point, the maximum point of extension of the water film occurs, further extending force resulting in breaking of the film and coalescence of the oil droplets. At this point a marked increase in stiffness is likely to occur, as the surface tension of the material in the bubble walls resists rupture. In mayonnaise this seems to occur at about 90% oil content.

Water content of many emulsions is critical. In mayonnaise, use of more than 15% water increases the probability that a separate aqueous layer will appear on long standing (Robinson 1924). Exposure of mayonnaise to drying atmospheres may cause coalescing of oil on the surface as the aqueous phase in the top layer is dehydrated. The ratio of emulsifying agent to water is critical, dilution of the emulsifier beyond its range of active concentration causing a breakdown of the system. This is seldom a problem in manufacture since the emulsifier is present in excess in most practical formulas. In fact, broken emulsions can often be reformed by adding the mayonnaise slowly to a small quantity of water while whipping. Gravies, sauces, and other emulsions stabilized with starch (or flour) are stable only within a rather narrow range of moisture content.

BUTTER

It is generally agreed that the qualities of the lipid components of butter are the major determinants of the product's texture, but several investigators have been able to show that butter texture is also affected by the distribution and form of the approximately 16% water normally found in it. Von Gavél (1956) studied microtome sections of fresh, cooled (32° to 37°F.) butter. The size of the water droplets was not constant but varied according to the availability of water. The solution in the droplets seemed to contain hygroscopic substances. Butter having a crumbly structure due to poor working exhibited additional (i.e., non-globular) water in crevices. Droplet size may very well influence texture in a manner analogous to the effect of bubble size on the firmness of whipped cream and meringues. In the latter products, increasing stiffness is observed as the bubbles become finer and more uniform.

Mulder (1953) indicated that normal differences in the moisture content of butter did not affect firmness. However, the degree of dispersion of the water may have had some effect on the hardness of his samples since he found that salted butter, which contains relatively large droplets, was softer than unsalted butter. Mohr and von Drachenfels (1956A) also discussed the effect of amount and distribution of water on consistency and spreadability. These authors (1956B and 1956C) proposed a scheme for evaluating the water distribution in butter depending on measurements of the dimensions of the water droplets. Class I (good) would be composed of those butters containing no droplets larger than ten microns in diameter. Class II (moderately good) butters could contain several droplets between 10 and 20 microns in diameter, while Class III (poor) products would include numerous droplets larger than ten microns, some being 30 microns or even larger in samples studied by Mohr and von Drachenfels.

CHOCOLATE

The moisture in chocolate is less than two per cent. There do not appear to have been any publications describing the effects of variations in water content on texture. However, one form of defect of appearance and texture called "bloom" is caused by a transient local excess of water.

There are two distinct types of "bloom" and they are unrelated in cause. Fat bloom results from the accumulation of cocoa butter on the surface of the piece. It has no relationship to moisture movement. Sugar bloom results from the formation of patches of sucrose crystals on the surface of sweet chocolate, milk chocolate, etc. The result is a granular and gritty texture and a rough, whitish appearance. Under normal conditions the sugar and other soluble materials in a piece of chocolate are completely immobilized by the absence of liquid water. However, wetting of the surface, due either to condensation of atmospheric water vapor on a chilled piece which has been exposed in warmer air or to accidental contact with liquid water, can dissolve the sugar (apparently to an appreciable depth) and allow it to concentrate on the surface. As the water again evaporates, the sugar is deposited as crystals of visible size.

HARD CANDIES

Hard candies are made by evaporating concentrated syrups of sucrose mixed with invert sugar and corn syrup. Less than two per cent water remains in the cooked mass at the time it is formed into pieces. In practice, the moisture content is controlled by terminating the cooking step at a definite temperature. Removal of water can be facilitated by cooking the syrup in vacuum pans, the lowered pressure allowing production of syrups having the desired water content at lower temperatures than would atmospheric pressure. The moisture content of the candy has a definite effect on its texture. If the moisture content is as high as four per cent, the product is noticeably softer in the mouth even though it is not chewed. Hardness increases rapidly as the moisture is brought down from the four per cent level. At about 1.5 per cent, a peak in hardness is reached which is not much affected by removal of additional water.

Heiss (1959) described an apparatus for measuring the important textural quality of surface stickiness. This instrument was intended to quantitate the evaluation obtained by pressing a "dry and clean finger" on to the candy momentarily and then pulling it away. In the Heiss apparatus, an upper plate cushioned with rubber is connected to a calibrated spring. A lower movable plate carries the candy piece which is made so that it is perfectly flat on the upper surface. The upper plate is initially pressed on to the candy with a weight of 11.5 oz. for ten seconds and then the lower plate is gradually withdrawn at a constant rate of speed until the candy surface and the rubber surface are pulled apart. The distance traveled before the release occurs is read off of a scale inscribed on the upright beam which carries the lower plate. The figure so obtained is a measure of the degree of stickiness of the candy surface. According to the data published by Heiss, the instrument gives reliable and reproducible results.

Hard candies are metastable systems which can maintain their characteristics only within a rather limited range of conditions. Because of their highly supersaturated status, they have a pronounced tendency to revert to the more stable crystalline form when conditions which allow the sugar molecules to move into an ordered array come into being. This movement is ordinarily restricted by the presence of intervening molecules of foreign species and by the limited availability of water. The later condition not only inhibits the mobility of the carbohydrate molecules but also limits the number of preferred hydrate crystal lattices which can form.

Although increasing the amounts of corn syrup carbohydrates and invert sugar solids tends to prevent crystallization by interfering with the formation of sucrose lattices, these ingredients also tend to make the product more hygroscopic so that, in atmospheres of equal relative humidity, larger quantites of water may become available to mobilize the sugar molecules.

The equilibrium relative humidities of hard candies varies with their composition. Figures of 25 to 40% relative humidity are frequently given as representative averages. The Heiss paper previously quoted contains extensive data on the response of hard candy of different composition to fluctuations in the ambient relative humidity (see Table 23) and the following discussion is based largely on that material.

TABLE 23

CHANGES IN HARD CANDIES STORED FOR 30 DAYS AT
70 PER CENT RELATIVE HUMIDITY[1]

Formula, 100 Parts Sucrose with:	Average Thickness of Grained Layer (Mm.)	Average Weight Increase (Per Cent)
90 parts of corn syrup[2]	~6.5	~1.35
65 parts of corn syrup	~5.0	~1.55
45 parts of corn syrup	~3.5	~1.8
15 to 20 parts of invert syrup	~5.5	>2.6
30 parts of invert syrup	~7.0	~1.65

[1] From Heiss (1959).
[2] Corn syrup of 43.9 dextrose equivalent was used in this experiment.

When hard candies are exposed to an atmosphere of higher relative humidity than their equilibrium value at the effective temperature, they will, of course, absorb water. The initial textural change is the development of a sticky surface as a result of the solution of the sugar that is in the layers near the surface of the candy. This stickiness will continue to increase, if the atmospheric relative humidity is sufficiently high, until enough water has been taken up to allow crystallization or "graining" of the surface layer.

Within certain ranges of relative humidity, candies of any given composition can develop stickiness without ever graining. For example, at 35% R.H., only those candies made with substantial quantities of added invert sugar will grain. At higher relative humidities, the stickiness period is shorter, but the degree of stickiness which is attained is more extreme.

Stickiness generally reaches a maximum twice in the deterioration process—once shortly before the beginning of crystallization and again when the crystal lattice begins to soften. The latter maximum represents a very advanced state of breakdown which is not usually seen under practical conditions of shortage. Stickiness tends to decrease with the onset of crystallization. Coating the pieces with granulated sugar, as is sometimes done, conceals but does not prevent undesirable texture changes.

To prevent deterioration due to the fluctuations in ambient relative humidity which are unavoidable during normal storage, handling, and display, it is essential to pack hard candies in containers having a low moisture-vapor transmission rate. Hermetically sealed tins or glass jars are, of course, ideal. In these containers, an atmosphere of favorable relative humidity is quickly established by the candy itself, and the product will persist unchanged for a very long time. Heat-sealed foil laminates are also very satisfactory. Wrapping of the individual pieces in waxed paper before packaging assists in retarding water vapor absorption.

SYRUPS

The viscosity of honey, molasses, maple syrup, corn syrup, and blends of sucrose and water with any of these, is recognized as a quality factor, high viscosity being equated with "richness" by most consumers. For any given solute composition, the viscosity is a function of the water concentration. Within the acceptable range of concentrations, the change of a per cent or so of water can cause perceptible changes in viscosity.

Fondant, although a very simple system, presents some interesting problems in texture and water relationships. Fondant is made by cooking a high proportion of sucrose, say 80%, with water to a temperature sufficient to completely dissolve all of the solid substance, and then cooling the solution with continual vigorous agitation so that a very fine crystal structure is formed. This material is used as a base for candies, icings, etc. Sometimes "doctors" such as corn syrups, sorbitol, or even egg white are added to inhibit crystal growth. A variation of as little as one per cent moisture can change the textural qualities of the product.

Upon standing for a day or so, the fondant preparation "ripens," i.e., it becomes softer, more plastic, and easier to work. Apparently, ripening is caused by a change in the crystal-size distribution, the smaller crystals disappearing and the larger crystals increasing slightly in size. The net effect of this change is that there is less crystal surface area available and the amount of fluid phase bound by the surface forces is less. A greater amount of the fluid becomes "free" and available to lubricate the crystals so that they move more freely on one another. Crystals continue to grow in size during storage but there is not much effect on texture of the additional growth, probably because the change in surface area is not as great once the tiniest crystals have disappeared. Growth of crystals can be prevented or at least inhibited by adding a small amount of a suitable hydrocolloid to the fondant. In this manner, diffusion of sugar molecules through the fluid phase is restricted. Some authorities have indicated that these doctors may also interfere directly with attachment of dissolved sugar molecules to the crystal surfaces by interposition.

GELS

Gels are commonly regarded as two-phase systems with a high degree of interface between a continuous, or at least intermeshed, system of solid material holding an aqueous (or other solvent) phase which may also be continuous or finely dispersed. Water is the only solvent of importance in food gels. The solid material is usually thought of as fibrillar in form and may consist either of strongly solvated molecules, such as proteins, or of thread-like crystals, as in gels of silicic acid or crystalline cellulose. A certain degree of structural stability is contributed by the extended skeleton of the continuous phase so that the system as a whole possesses many of the properties of solids.

The amount of water present in a gel is not a fixed quantity. If the gel is placed in an excess of water it will usually swell until a sol is formed. On the other hand, solvent can be removed by thermal methods or by applying pressures. The forces required to squeeze the water out of gels are usually quite large. Conversely, dehydrated gels can take up water against very high pressures. However, solvent is often spontaneously expressed from gels as they age. This process, called syneresis, is evidently due to the formation of additional intermolecular bonds with a consequent reduction in the number of loci available for solvent binding and to a decrease in the dimensions of the intermolecular spaces in which the solvent is contained. Change of the system from a viscous sol to a firm gel may occur over a narrow range of water content, as is the case with agar, or the transition may be gradual.

Addition to the sol of molecules which compete with water for bond sites profoundly affects the quality of the gel, usually in the direction of increased rigidity. Similar effects occur when ions capable of bridging electrophilic or electronegative sites on neighboring molecules are included in the system. Calcium is particularly effective in this regard if the gel contains a considerable number of carboxyl groups.

Fruit jellies are based on pectin. This substance, which is widely distributed in nature but is of principal importance in fruits, disperses in water to form a viscous colloidal sol. The property of most importance to the student of food texture, however, is its ability to form sugar-acid-pectin gels, i.e., the fruit jellies, jams, and preserves of commerce. Gel formation of pectin with sugar and acid was shown by Olsen (1934) to be in accord with the following assumptions:

(1) pectin is a negative hydrophilic colloid.

(2) the sugar functions as a dehydrating agent.

(3) hydrogen ions function by reducing the negative charge on the pectin, thus permitting coalescence of the molecules to form a network.

(4) dehydration of the pectin requires time to come to an equilibrium, so that some gels set more quickly than others or require different setting temperatures.

(5) the rate of dehydration and precipitation increases directly as the hydrogen ion concentration increases.

(6) maximum jelly strength is reached at equilibrium.

Although water content is one of the principal determinants of texture, other factors are also important in fruit jellies.

Cheese is representative of a type of food gel. Texture attributes are strongly related to the moisture content. Soft cheese varieties may contain from 40 to 75% moisture while hard cheese will average between 30 and 40% water. In freshly made Cheddar cheese, the fat-free portion will contain about 55 to 54%. Processed cheese contains added water, partly for the economic advantage, to be sure, but also as a means of softening the product to create a texture that is desired by the majority of consumers. Low moisture content in natural cheese may contribute to "corky" texture, while high moisture is sometimes the cause of a mealy or pasty condition.

Marshmallows

Marshmallows are examples of a foam-type product having a texture affected noticeably by small changes in moisture content. They are usually composed of sugars, water, gelatin, and flavoring. A complete analysis of the fundamental physical factors contributing to perceived texture in marshmallow, and their relationship to processing variables has been published by Tiemstra (1964A, 1964B, 1964C, and 1964D). The moisture content of marshmallow will vary throughout the manufacturing process as well as during the shelf-life of the product.

The ratio of gelatin to moisture, i.e., the gelatin solution concentration, is directly related to the elasticity modulus for any grade of gelatin as measured by gel strength. Both sugar composition and moisture level have major influences on viscosity. The gelatin solids have to be balanced with the moisture content of the final marshmallow piece to obtain good eating qualities and still give resilience and ability to withstand handling.

The overrun or amount of air whipped into the gelatin is, of course, related to texture in that the apparent resistance to chewing of a given volume of product (i.e. pieces of uniform size) is less with increased overrun. The overrun obtained from a marshmallow syrup under identical conditions varied inversely with the viscosity of the syrup and directly with the moisture content. Other factors influencing the overrun such as the temperature of optimum whip change with variations in moisture content.

BIBLIOGRAPHY

BARRETT, C. D. 1961. Temperature and relative humidity determine candy shelf life. Food Processing 22, No. 4, 46–47.

BRADLEY, W. B. 1949. Bread softness and bread quality. Bakers Digest 23, No. 1, 5–7.

CLELAND, J. E., and FETZER, W. R. 1944. Moisture absorptive power of starch hydrolyzates. Ind. Eng. Chem. 36, 552–555.

COOK, H. T., PARSONS, C. S., and McCOLLOCH, L. P. 1958. Methods to extend storage of fresh vegetables aboard ships of the U.S. Navy. Food Technol. 12, 548–550.

CRAMER, A. B. 1950. Some problems in the manufacture of hard candy. Food Technol. 4, 400–403.

DITTMAR, J. H. 1935. Hygroscopicity of sugars and sugar mixtures. Ind. Eng. Chem. 27, 333–335.

DUCK, W. 1959. Consistency of caramel. Mfg. Confectioner 39, No. 6. 29–31.

GRAU, R., and HAMM, R. 1953. A simple method for determining water-binding in muscle. Naturwissenschaften 40, 29–30.

GROVER, D. W. 1947. The keeping properties of confectionery as influenced by its water vapor pressure. J. Soc. Chem. Ind. London 66, 201–205.

GROVER, D. W., and NICOL, J. M. 1940. The vapor pressure of glycerol solutions at 20°C. J. Soc. Chem. Ind. 59, 175–177.

HEISS, R. 1959. Prevention of stickiness and graining in stored hard candies. Food Technol. 13, 433–440.

HUELSEN, W. A. 1954. Sweet Corn. Interscience Publishers, New York.

JORDAN, S. 1926. Chemistry and confectionery. Ind. Eng. Chem. 16, 336.

KARÁCSONY, D., and PENTZ, L. 1955A. Further theoretical and practical investigations on the candying and storage of fondant. Élelmézesi Ipar 9, 236–244.

KARÁCSONY, D., and PENTZ, L. 1955B. Theoretical and practical problems on the candying and storage of fondant. Élelmézesi Ipar 9, 45–52.

LIPSCOMB, A. G. 1956. Rheology of candy syrup and boiled candies. Fette Seifen Anstrichmittel 58, 875–879.

LOWE, B. 1955. Experimental Cookery from the Chemical and Physical Standpoint. 4th Edition. John Wiley and Sons, New York.

MACZELKA, L. 1956. Colloid aspects of food chemistry and technology. Élelmézesi Ipar 9, 199–202.

MAKOWER, B., and DYE, W. B. 1956. Equilibrium moisture content and crystallization of amorphous sucrose and glucose. J. Agr. Food Chem. 4, 72–77.

MALLOWS, J. H. 1960. Foods of simple structure. Soc. Chem. Ind. Monograph 7, 10–13.

MATZ, S. A. 1962. Food Texture. Avi Publishing Co., Westport, Conn.

MOHR, W., and VON DRACHENFELS, H. J. 1956A. Consistency of butter and margarine. Fette Seifen Anstrichmittel 58, 609–613.

MOHR, W., and VON DRACHENFELS, H. J. 1956B. The microscopic determination of the water distribution in butter, a method for the recognition of the correct working and forming procedure of butter. Milchwissenschaft 11, 126–132.

MOHR, W., and VON DRACHENFELS, H. J. 1956C. Observations on frozen sections of butter as a method for the study of the disposition of crystals and distribution of water in butter. Milchwissenschaft 11, 228–234.

MULDER, H. 1953. The consistency of butter. In Foodstuffs: Their Plasticity, Fluidity, and Consistency. Edited by G. W. Scott Blair. Interscience Publishers, New York.

MULLIN, J. W. 1961. Crystallization. Butterworth, London.

NEVILLE, H. A., EASTON, N. R., and BARTRON, L. R. 1950. The problem of chocolate bloom. Food Technol. 4, 439–441.

OLSEN, A. G. 1934. Pectin studies. III. General theory of pectin jelly formation. J. Phys. Chem. 38, 919–930.

PALMER, K. J., DYE, W. B., and BLACK, D. 1956. X-ray diffractometer and microscopic investigation of crystallization of amorphous sucrose. J. Agr. Food Chem. 4, 77–81.

PARSONS, C. S. 1959. Effects of temperature and packaging on quality of stored cabbage. Proc. Am. Soc. Hort. Sci. 74, 616–621.

PARSONS, C. S., and WRIGHT, R. C. 1956. Effects of temperature, trimming, and packaging methods on lettuce deterioration. Proc. Am. Soc. Hort. Sci. 68, 283–287.

ROBINSON, S. K. 1924. Practice in mayonnaise manufacture. Am. Food J. 19, 185–190.

SALWIN, H. 1959. Defining minimum moisture contents for dehydrated foods. Food Technol. 13, 594–595.

SALWIN, H., and SLAWSON, V. 1959. Moisture transfer in combinations of dehydrated foods. Food Technol. 13, 715–718.

SHEPPARD, S. E., and SWEET, S. S. 1921. The elastic properties of gelatin jellies. Am. Chem. Soc. 43, 539–547.

SOKOLOVSKY, A. 1937. Effects of humidity on hygroscopic properties of sugars and caramels. Ind. Eng. Chem. 29, 1422–1423.

SWIFT, C. E. WEIR, C. E., and HANKINS, O. G. 1954. The effect of variations in moisture and fat content on the juiciness and tenderness of bologna. Food Technol. 8, 339–340.

THIEME, J. G. 1934. The hygroscopic properties of raw sugar and molasses. Arch. Suikerind. 42, 157–180.

TIEMSTRA, P. J. 1964A. Marshmallows. I. Overrun. Food Technol. 18, 915–920.

TIEMSTRA, P. J. 1964B. Marshmallows. II. Viscosity and elasticity. Food Technol. *18*, 921–927.

TIEMSTRA, P. J. 1964C. Marshmallows. III. Moisture. Food Technol. *18*, 1084–1091.

TIEMSTRA, P. J. 1964D. Marshmallows. IV. Set and syneresis. Food Technol. *18*, 1091–1096.

VON GAVÉL, L. 1956. The structure of butter on the basis of observations on microtome sections. Z. Lebensm. Untersuch. u. Forsch. *104*, 1–21.

WECKEL, K. G., SCHARSCHMIDT, R. K., and RIEMAN, G. H. 1959. Sloughing in canned potatoes. Food Technol. *13*, 456–459.

WHITTIER, E. O., and GOULD, S. P. 1930. Vapor pressures of saturated equilibrated solutions of lactose, sucrose, glucose, and galactose. Ind. Eng. Chem. *22*, 77–78.

WILSON, G. D. 1960. Sausage products. *In* The Science of Meat and Meat Products. W. H. Freeman and Co., San Francisco.

WISMER-PEDERSEN, J. 1959. Quality of pork in relation to rate of pH change post mortem. Food Research *24*, 711–727.

CHAPTER 17

Moisture Transfer in Finished Products

INTRODUCTION

Net movement of moisture between different components in the same package or between foodstuffs and the atmosphere is frequently responsible for deleterious changes during storage, transfer, and display of the products. This movement may be the result of liquid transfer under gravitational or capillary forces, but probably more often is the result of translocation through the vapor state.

MOISTURE TRANSFER BETWEEN COMPONENTS IN A PACKAGE

Non-homogeneous foods whose particles are in contact with a common vapor phase in a package made of essentially impermeable material will eventually reach a state in which the equilibrium relative humidity (or water activity) of each component is the same, assuming constant temperature and pressure and the absence of microbiological or chemical actions which give off or consume water. The smaller the particles and the closer the contact (i.e., the shorter the path of the water molecules), the faster equilibrium will be attained. Interchange of this sort is of practical importance in combination packs of dehydrated foods, in jelly- or marshmallow-filled cookies, etc.

The Armed Forces and other organizations utilize dehydrated foods such as stews which are formed by mixing separately dried components. The mixture is packed in a plastic-and-foil laminated pouch to prevent interchange of moisture with the atmosphere. Before they are mixed, each of these components will ordinarily have a different moisture content and a different equilibrium relative humidity. After they are mixed together, water molecules will pass between the various materials until there is finally reached a moisture content in each which will support the same equilibrium relative humidity. At this equilibrium point, some of the sensitive components may have attained water activities which are incompatible with stability requirements even though their moisture contents when packed were well below the safe maximum. For example, in ground beef and potato hash (Salwin 1963) the equilibrium relative humidity of the beef is about two per cent, while the figures for onions, potato cubes, and the gravy base range from 20 to 35%. In such cases, it is important to be able to compute the moisture content that will be reached by each component when equilibrium has been established.

235

As shown by Salwin (1959) and Salwin and Slawson (1959), the equilibrium moisture vapor pressure (R) of a combination of dehydrated foods can be calculated from the equation.

$$R = \frac{R_a S_a W_a + R_b S_b W_b}{S_a W_a + S_b W_b}$$

when W_a and W_b are the dry weights of the components, R_a and R_b are the initial relative humidities of the components, and S_a and S_b are the slopes of the isotherms between the initial and final points. A typical application of this equation is given in Table 24. If it is found that a combi-

TABLE 24

PREDICTED AND OBSERVED MOISTURE TRANSFER IN PRECOOKED DEHYDRATED CHICKEN STEW STORED AT 72 °F.[1]

Ingredient	Weight, Grams	Relative Humidity, Per Cent		Moisture, Per Cent of Solids		
		Initial	Final Predicted	Initial	Final Predicted	Observed
Chicken	39.9	1.1[2]	7.6	1.07	3.17	3.18
Potato dice	30.1	28.4	7.6	5.88	3.45	3.64
Lima beans	15.2	7.7	7.6	3.68	3.68	3.90
Cream sauce base	5.0	18.8	7.6	3.49	2.60	2.54
Chicken soup and gravy base	4.0	37.0	7.6	2.49	1.17	1.19
Mixture, calculated total moisture (grams)				2.90	2.94	3.03

[1] Salwin (1963).
[2] Extrapolated value.

nation pack of two or more components would be impractical because of the difficulty of reducing the moisture content of a major constituent to a compatible level, separate containers will have to be provided.

MOISTURE TRANSFER BETWEEN PRODUCT AND ATMOSPHERE

It is usually desirable to minimize the interchange of moisture vapor between atmosphere and product. In some cases, as with certain fresh fruits and vegetables, outlets for moisture given off by the product must be provided so that droplets do not condense on the packaging material. Generally, however, processed foods are packaged at about their optimum moisture content. Uptake or loss of water can be deleterious to quality. Products which may become unsatisfactory as a result of decreases in moisture content are pop corn (unpopped), which will not pop below a certain moisture level, dried fruit, which becomes leathery and tough as it loses water, cheese, and meat. Freezer burn, an unsightly but harmless condition occasionally seen in frozen foods stored for long times, is due to sublimation of moisture from a localized area of the surface tissue and can be prevented by packaging in protective films. Foods which lose acceptability or become susceptible to microbiological spoilage, as a result of absorption of water vapor from the atmosphere are hard candy, many breakfast cereals, fried snack items (potato chips, bacon rinds, corn curls, etc.), caramel corn, and most dehydrated foods.

Development of containers which will prevent exchange of moisture vapor has engaged the attention of packaging engineers for many years.

Resistance of any package to moisture vapor transfer is affected by three factors: (1) the permeability of the paper, board, or film making up the package; (2) the efficiency of the heat seal or other closure; and (3) the mechanical strength of the wrap, i.e., the extent of loss of barrier properties at bends, creases, folds, etc. However, the inherent permeability of the film is usually the controlling property. In selecting material for a container, factors such as cost, sealing characteristics, strength, transparency, and printability must also be taken into consideration so that a compromise with water vapor permeability may be necessary. Some extremely sensitive foods, mostly very low moisture dehydrated foods, are still being packaged in cans, especially in institutional sizes, but this is partly to obtain protection against mechanical damage.

Permeabilities of intact packaging films vary according to the inherent properties of the material, the area exposed, the temperature, the difference in water activities on the opposite sides of the film, the thickness of the film, and the pressure. The quantitative effects of changes in these factors on the water vapor transmission rate cannot always be predicted. Consequently, it is the usual practice to test under conditions representing the extremes expected to be encountered in actual storage and to reserve extrapolation for the time factor.

There are three common methods which can be used to determine the permeability of a film to gases. These are measurements of (1) volume increase; (2) pressure increase; and (3) concentration increase. In the case of water vapor permeability, all of the techniques which have seen widespread use involve the collection by a desiccant of the moisture passing through the film. Measurement of water-vapor transmission rate according to Method E-96 of the American Society of Testing Materials is based on this principle. A test chamber in which the relative humidity and temperature can be closely controlled is required. Moisture vapor from the test chamber permeates a measured area of the sheet or film which has been sealed over the mouth of a cup containing the desiccant. Gain in weight by the desiccant over a measured period of time is accepted as the amount of moisture vapor passing through the film. The data are reported as grams of water which will pass through a standard area, usually one square meter but occasionally 100 sq. in., in a given time, usually 24 hr. Conditions of 90% relative humidity and 100°F. in the test chamber are standard. In all following discussions, data are reported on the basis of one square meter of film. The desiccant to be used is prescribed for standard tests. Table 25 includes data describing the effectiveness of various drying agents.

TABLE 25

EFFECTIVENESS OF VARIOUS DRYING AGENTS[1]

Drying Agent	Residual Moisture, Milligrams of Water Per Liter of Air
Cooling to $-21°$ (salt-ice mixture)	4.5×10^{-2}
Cooling to $-72°$ (carbon dioxide snow)	1.6×10^{-2}
Cooling to $-194°$	1.6×10^{-23}
Phosphorus pentoxide, P_2O_5	$<2 \times 10^{-5}$
Magnesium perchlorate, "Anhydrone," $Mg(ClO_4)_2$	$<5 \times 10^{-4}$
Magnesium perchlorate trihydrate, "Dehydrite," $Mg(ClO_4)_2 \cdot 3H_2O$	$<2 \times 10^{-3}$
KOH (fused)	0.002
Aluminum trioxide, activated alumina, Al_2O_3	0.003
Sulfuric acid	0.003
Calcium sulfate, "Drierite," $CaSO_4$	0.004
Magnesium oxide, MgO	0.008
Calcium chloride, $CaCl_2$	0.14 to 0.25

[1] Adapted from Critical Tables and Morton (1938).

Brickman (1957) discussed the engineering approach to establishing the package specifications for a product. He took into consideration two distinct levels of moisture content: (1) the normal moisture content, which contributes to good storage stability; and (2) the failure-point moisture content, or the level at which the product becomes unacceptable to the consumer. Although the methods for determining these two levels are perhaps lacking in precision, Brickman's approach is valuable as a quantitative basis for the evaluation of the commercial value of a packaging material. The procedure involves (1) plotting the "equilibrium per cent moisture" versus "per cent relative humidity" at a given temperature; (2) calculating the amount of water that would have to pass through the walls of a package in order to change the moisture content from the normal to the failure-point level; and (3) choosing for the test those packaging materials which have WVTR within a range adequate to provide the desired shelf life. When promising materials have been selected, samples of the product are sealed in packages made of them, and held in storage for 90 days under controlled conditions of temperature and relative humidity. The rate of change of moisture content in the sample is determined by periodic weighings.

The weight change, expressed as per cent of the original weight of the loaded bag or pouch, is plotted against days in storage for each combination of storage temperature and relative humidity. A more-or-less linear plot is obtained for each packaging material. From this curve the time required to reach the failure point-moisture content can be estimated for each set of conditions.

Brickman's original method was recently modified (Brickman 1961) to eliminate some special equipment and to simplify the obtaining of permeability measurements over a wide range of temperature and humidity conditions. A temperature range of 0° to 100°F. and a humidity range of 30 to 95% was suggested for use in the new "pouch" method. The data are plotted as WVTR vs. the reciprocal of the absolute temperature on semilog paper, supposedly to conform to the Arrhenius relationship.

Ball (1963) took exception to the method of graphing data suggested by Brickman on the grounds that it was theoretically unsound. He preferred to plot time against the difference between the per cent of moisture in the packaged product and the equilibrium per cent moisture content, i.e., D – E using his symbols. The resulting curves should be straight lines on semilog paper and can validly be extrapolated beyond the 90-day test period.

Papers and Paperboards

The papers most commonly used for packaging foodstuffs are kraft, glassine, greaseproof, and waxed. Uncoated glassine has no useful moistureproofness unless it is coated or laminated with more resistant materials. Lacquered and waxed glassines can exhibit WVTR as low as 3.0 gm./24 hr./sq. meter, which makes them suitable for many purposes where moderate protection is required. The barrier properties of laminated glassines depend entirely on the nature of the laminant and the number of plies.

Kraft-, pouch-, extensible-, and greaseproof-paper, like glassine, have little resistance to the transfer of moisture vapor. Polyethylene coated-kraft paper and waxed paper have fairly low WVTR, in some cases approaching 5.0 gm./24 hr./sq. meter if the thickness of the base paper and amount of coating are adequate. Since these barrier properties depend so much on characteristics of the coatings, they will be discussed in a later section devoted to coatings.

Film and Foil

Intact aluminum foil is, as would be expected, impervious to water vapor. However, the development of pinholes during manufacture or during subsequent handling may reduce the protective value of the material. Table 26 indicates the rate of occurrence of pinholes in one series

TABLE 26

PINHOLES IN ALUMINUM FOIL AND THEIR RELATION TO WATER VAPOR TRANSMISSION[1]

Gage of Foil, In.	Sheets with Pinholes, Per Cent	Average WVTR, Gm./Sq. Meter/24 Hr.
0.00035	100	4.5
0.0005	100	1.9
0.0007	15	0.67
0.001	8	0.11

[1] Adapted from Anon. (1953).

of samples of commercial foil, and the effect of these pinholes on moisture vapor transfer. Lamination of the thinner foils to plastics improves their barrier quality considerably by reducing the effect of pre-existing pinholes and by increasing the resistance of the material to damage by creasing, scuffing, etc. By definition, aluminum foil is not thicker than 0.0059 gage. Gages 0.006 and higher are called sheet.

There are a large number of plastic films available for use as wraps contacting foods or as overwraps for food packages. Several of these have moisture vapor transmission rates almost as low as aluminum foil. Coated cellophanes, saran, and the higher density polyolefins provide the best levels of barrier properties against water transfer. Plain cellophane, cellulose acetate, most vinyl films, and some of the heavily plasticized Pliofilm grades have rates which are so high they cannot be regarded as effective moisture barriers.

Cellulose acetate, a thermoplastic film, can be made in forms suitable for direct contact with food. It has a high rate of water-vapor transmission, making it suitable for packaging fresh fruit and some bakery foods where fogging is a problem, but not for those products sensitive to moisture exchange.

Commercial polyethylene, a very common food packaging film, can be classified as low density (0.910–0.925 gm./ml.), medium density (0.926–0.940 gm./ml.), or high density (0.941–0.965 gm./ml.). Each of these types has distinct properties. The low cost and high yield of polyethylene film rather than any superior protective property is responsible for its popularity in the food field. The thinner grades of the low density material do not provide top protection against loss or absorption of moisture. Thicker layers of high density polyethylene give protection which is entirely adequate for most purposes.

Polypropylene films exhibit the highest tensile strength of any of the polyolefin films and have better impact strength and tear resistance than either the medium or higher density types of polyethylene at 72°F. The stiffness and abrasion resistance of the polypropylene films are comparable with the highest density polyethylene. They display a WVTR in the range of a medium-density polyethylene film, and offer gas barrier properties better than those of any of the three types of polyethylene films.

Polyester films (brand names Mylar, 3M, and Kodar) of differing composition are available. Mylar is a polymer of ethylene glycol and terephthalic acid. It can be used in contact with foods provided processing temperatures after sealing do not exceed 250°F. Good protection against liquid or vapor forms of water can be achieved by coating combinations or with proper film formulations.

The name Saran is applied to films made from a thermoplastic copolymer of vinylidene chloride and vinyl chloride. The resin is extruded in the form of a tube which is extended in diameter by air pressure and then immediately quenched in a cold water bath to keep it in an amorphous state. The process yields an oriented, highly flexible film which is both strong and tough. Saran film is produced in types identified by numbers 1 through 39. The first 19 members of this series contain the highest percentage of vinylidene chloride. They have superior barrier properties and chemical resistance. Formulations 20 through 39 are more flexible at low temperatures. Both types have very low water vapor transmission rates. Saran has been used rather widely for food packaging, especially for shrink wraps. It is also the principal film used for club packaging of cookie doughs, etc.

The water vapor transmission rate of oriented polystyrene film is relatively high but considerably lower than that of cellulose acetate. These films are biaxially oriented.

Pliofilm (rubber hydrochloride) has been used for packaging baked goods, cheese cuts, and meat. It has relatively good barrier properties against water vapor.

Various organic polymers containing fluorine have extremely low rates of moisture vapor transmission but there is some question about their use in contact with foods. Coatings of these compounds applied to the exterior side of wrapping materials are apparently fully acceptable. The cost is quite high at this time but some reduction may be anticipated if a volume outlet opens up.

Vinyl chloride can be polymerized in an aqueous solution of mixed oxidizing and reducing agents to yield polyvinyl chloride, a useful raw material for packaging films. The properties of the films are dependent on the plasticizers added, to a large extent. Some plasticizer combinations can be used with foods, others cannot. In thicknesses of about one mil, vinyl films can be expected to have MVTR in the range of 4 to 13. It has been used for shrink wrapping of fruits and vegetables, its relatively high MVTR being advantageous for these purposes.

Cryovac is the trade name of W. R. Grace and Co. for several plastic films. Cryovac Y is an oriented polypropylene which can be used for both shrink and non-shrink packaging. It has a MVTR of 3.1 gm./ml./sq. meter/24 hr. Cryovac D is a cross-linked medium density polyethylene especially designed for use on automatic soft film-overwrap equipment. Cryovac L is a cross-linked low density polyethylene for shrink cover packaging with an MVTR of 10 gm./ml./sq. meter. All of these are useful in food applications.

Permeability factors for films for general use are given in Table 27, while the rates for shrink pack films are tabulated in the subsequent table.

TABLE 27

PERMEABILITY OF SOME COMMON PACKAGING FILMS[1]

Film	Relative Permeability at 104 °F. to Water Vapor at 90 Per Cent R.H.
Polyesters (e.g., Mylar)	18.0
Low-density polyethylene	22.5
High-density polyethylene	6.7
Polypropylene	8.7
Cellulose (uncoated)	1865.
Cellulose (vinylidene chloride coated)	7.1
Polyvinyl chloride	27.3

Sherwood (1964).

TABLE 28

WATER VAPOR TRANSMISSION RATES OF SHRINK FILMS[1]

Film	WVTR[2]
Polyester	45
Polyethylene:	
Regular[3]	12–18
Crosslinked, Type I	10
Crosslinked, Type II	5
Polypropylene	4
Polystyrene	Over 60
Polyvinyl chloride	Over 50
Polyvinylidene chloride copolymer	3–20
Rubber hydrochloride	12–20

[1] Lowry 1964. Used by permission of the McGraw-Hill Publishing Co.
[2] ASTM Method E96-53T. Units are gm./sq. meter/24 hrs./mil at 100°F. and 90% R.H.
[3] Properties of regular polyethylene can vary with resin density or blend and manufacturing methods. Cross linked type I is low density and Crosslinked type II is medium density.

Coatings

Paraffin is regarded as nontoxic when used in accordance with FDA regulations and it provides moderately good protection against water vapor transfer when it is used as a coating on paper or paperboard (Wolper 1962). Although it is inexpensive and will form a relatively impervious seal, the seal is not strong and the coating is easily scuffed, leading to loss of some of its WVT protection. Modified paraffin coatings with up to ten per cent of microcrystalline waxes or certain resins show improved heat seal strength and film durability. When more than ten per cent of modifiers are added to a paraffin base, the combined material is usually referred to as a "hot-melt coating." These mixtures require sealer operating temperatures of 250° to 400°F. They yield strong heat seals but provide low protection against water vapor transmission.

Lacquer coatings can be classified broadly by the type of solvent employed as (1) organosols, in which a resin and a plasticizer are dispersed in a combination of polar and non-polar organic solvents, and (2) plastisols, in which synthetic elastomers are dispersed in liquid plasticizers.

TABLE 29

WATER VAPOR TRANSMISSION RATES OF SOME COMMON PACKAGING COATINGS[1]

Coating Material	Base Material[2]	Wt., Lb./3,000 Sq. Ft. (Reams)			Heat-Seal Range, °F. (Reciprocating Jaw, 0.5-Sec. Dwell)	WVTR[3]
		Coating	Base	Total		
Wax coatings						
Refined paraffin	MF sulfite	5	25	30	. . .	High
wax	SC sulfite	10	25	35	150–180	11–31
	Glassine	4	25	29	150–180	3–6
	Glassine	6	25	31	150–180	1.6–3.1
Modified paraffin	SC sulfite	10	25	35	150–180	11–31
wax	Glassine	8	25	33	150–180	1.6–3.1
Hot-melt types	SC sulfite	10	25	35	200–350	High
Solvent coatings						
Cellulose nitrate	SC kraft	8	32	40	200–300	19–28
	Glassine	3	22	25	200–300	6–12
	Cellophane	2	20	22	200–300	6–16
Cyclized rubber	SC kraft	8	32	40	200–300	9–16
	Glassine	3	22	25	200–300	2.3–3.9
Butadiene-styrene	Glassine	3	22	25	200–300	2.3–3.9
Vinyl copolymers	SC kraft	5	25	30	220–280	60–90
	Glassine	3	22	25	220–280	39–60
Vinylidene	SC kraft	6	25	31	250–300	31–46
chloride	Glassine	4	25	29	250–300	16–31
(saran-type)	Cellophane	4	20	24	250–300	8–11
Extruded coatings						
Polyethylene						
Low density						
0.5 mil	MF or SC kraft	8	25	33	200–300	39–54
1.0 mil	MF or SC kraft	15	25	40	200–300	19–25
2.0 mil	MF or SC kraft	30	25	55	200–300	9–12
0.5 mil	Foil-glassine	8	42	50	200–300	Less than 0.3
2.0 mil	Cellophane	30	24	54	200–300	8–11
Med. density						
1.0 mil	SC kraft	15	25	40	250–350	16–22
High density						
1.0 mil	SC kraft	15	30	40	275–375	9–16

[1] Adapted from Modern Packaging Encyclopedia issue for 1965. By permission of Packaging Catalog Co. and McGraw-Hill Book Co.
[2] MF = machine finish; SC = supercalendered.
[3] Water vapor permeability is given in gm./24 hr./sq. meter at 100°F. and 90% relative humidity.

Such coatings, including rubber, butadiene-styrene, copolymers, and Saran compounds provide good protection against the passage of moisture in vapor or liquid form. Vinyl copolymers are intermediate in this respect, while the cellulose derivates are not very effective.

Polyethylene, polypropylene, Saran (polyvinylidene chloride), and nylon (polyamide) form the extrusion category of coating materials. The worth of polyethylene as a protective layer for application to other packaging materials is strongly dependent on the thickness of the film and the density of the plastic, the thinner layers of low-density polyethylene having little value. Nylon has somewhat similar attributes while Saran is an excellent barrier material. A summary of the water vapor transfer rates of the more common coating substances is given in Table 29.

SOME EXAMPLES OF PACKAGING PRACTICES

Waxed papers and cellophane are the traditional wraps for bakery foods, but in recent years coated and laminated polyolefins have taken over large segments of the market, and especially for sliced loaves. As indicated in Table 30, the moisture permeability rates of the polyolefins in

TABLE 30

WATER VAPOR PERMEABILITY OF SOME TYPICAL BREAD-WRAP FILMS[1]

Film	Water Vapor Permeability[2]
Cellophane (250 polymer)	9
Cellophane (220 nitrocellulose)	16
Cast polyethylene	12
Cast polypropylene	9
Oriented polypropylene (unbalanced)	7
Modified polyolefin[3]	7
Modified polyolefin[4]	11

[1] Davis (1964).
[2] Grams/24 hr./mil/meter at 100°F. and 90 per cent R.H.
[3] A three-ply laminate with oriented polypropylene center and polyethylene outer layers.
[4] A three-ply laminate with an ultra-thin polypropylene center and polyethylene outer layers.

useful gages are lower than those of waxed paper and cellophane. As a practical matter, staling due to drying has not been an important problem with any of these wraping materials and the change-over has been motivated by other considerations.

For cookie packaging, Saran latex coating is coming into use. Here, protection against moisture absorption is the important factor. Powdered milk requires good protection, such as 0.0035 aluminum foil laminated to polyethylene, waxes, and papers to get WVTR of 0.02 to 0.04 gm./24 hr./100 in². Most dehydrated foods such as potato granules require barrier material of similar effectiveness but potato flakes and slices can apparently be successfully packaged in coated folding cartons of food board if the seal is impervious (protective carton overwraps are often used).

Cellophane

Cuts of fresh meat will turn dark if they are permitted to dehydrate in open air. Hygroscopic packaging materials can cause surface desiccation.

Cellophane, properly modified, has been successfully applied to the packaging of self-service meat. This modification of cellophane is coated on one side with a nitrocellulose or polymer coating which is highly permeable to oxygen but impermeable to moisture vapor. The uncoated side is placed in contact with the meat surface. Since uncoated cellophane is wettable, liquid from the meat soon saturates the film up to the coating. The total amount of moisture taken up in this manner is small and does not affect the surface appearance of the meat. The oxygen permeability of the wet cellophane is much greater than that of the dry material, resulting in longer retention of the bright red color of beef, etc. In this way, to the desirable properties of cellophane are added the characteristics of low moisture transfer and high oxygen permeability, both of which are highly desirable for fresh meat. Elimination of air pockets as a result of adherence of the film to the meat surface increases the brilliance of appearance.

Ball (1957 *et al.*) determined weight loss undergone by samples of meat packaged in films and stored for three weeks. This loss, presumably due to evaporation of moisture from the meat, was held to less than one to two per cent by (1) cans, (2) cellophane-pliofilm laminate with pliofilm inside, (3) cellophane-polyethylene laminate with polyethylene inside, and (4) cellulose acetate-pliofilm laminate with pliofilm inside. Losses of about three to four per cent were observed in samples packaged in MSAT86 cellophane, coated both sides. Fair protection (ten per cent decrease in weight) was afforded by MSAT80 cellophane, coated both sides. The MSAT cellophanes had thicknesses of about 1.1 mils. The cellophane-pliofilm laminate was 2.74 mils thick while that of the cellophane-polyethylene laminate was 3.29 mils. The cellulose acetate was 2.43 mils in thickness.

Controlled Storage

When humidity-sensitive raw materials are to be stored for long times in bulk, the use of protective packaging is often impractical. A simple alternative is storage under conditions of controlled humidity and temperature. Similar expedients are sometimes desirable for stocks of finished products which must be held for lengthy periods before they can be distributed. Many packaging materials also take on undesirable qualities if they are allowed to dry out or absorb relatively large amounts of moisture. Several authors (e.g., Zenlea 1934; Nelson 1949; Ball 1963; and Wright *et al.* 1954) have published information on optimum storage conditions for raw

TABLE 31

OPTIMUM CONDITIONS FOR STORAGE OF INGREDIENTS, PACKAGING MATERIALS, AND
NON-FROZEN FOODS[1]

	Storage Conditions	
Product	Temperature, °F.	Relative Humidity, Per Cent
Avocados	37–48	85–90
Bacon, farm style	60–65	85
Bacon, commercial	34–40	85
Bananas	53–60	85–90
Beans, fresh green or lima	35–45	85–90
Candy, chocolate covered	35	40–50
Candy, hard	60–80	40–50
Cheese, Cheddar and similar types	35	65–70
Cocoa	60–70	50–65
Coffee, green	35–37	80–85
Corn, sweet	31–32	85–90
Cucumbers	45–50	90–95
Eggplants	40–45	85–90
Eggs, shell	29–30	85–92
Eggs, dried products	35	10
Fiber boxes	70	50
Fish, fresh	32–35	90–95
Fruits, dried	32	50–60
Grapefruit	50–60	85–90
Mangos	50	85–90
Marshmallows	60–80	65
Meats, fresh cuts (pork, beef, lamb)	32–34	85–90
Melons, ripe		
Watermelons	36–40	75–85
Others	40–50	75–85
Milk, dried products	40	10
Nuts	32–36	65–70
Okra	50	85–95
Olives, fresh	45–50	85–90
Oranges, California	35–37	85–90
Oranges, Florida	30–32	85–90
Paper, wax	45–80	40–60
Pecans	35	90–92
Peppers, green	45	85–90
Pineapples		
Mature green	50–60	85–90
Ripe	40–45	85–90
Potatoes, white		
For general use (baking, etc.)	38–40	80–90
For chipping or frying	50–55	80–90
Poultry, fresh	32	85–90
Pumpkins	50–55	70–75
Sausage, fresh pork	32	85–95
Squash, summer	40–50	85–95
Squash, winter	50–55	70–75
Sugar, brown	<75	60–70
Sugar, white	<100	<60
Sweet potatoes and yams, cured	55–60	80–85

[1] See text for sources.

materials, packing materials, and finished products. A compilation of these data is given in Table 31.

Most fresh root, leafy, and stalk vegetables should be stored at temperatures slightly above 32°F. and at high relative humidity, preferably about 90 to 95%. Berries, cherries, grapes, peaches, plums, pomegranates, apples, and many other fruits retain acceptable quality longest at about 31° to 32°F. and relative humidities of 85 to 90%. These humidities are sufficiently high to prevent excessive loss of moisture from the product but are not high enough to promote rapid mold growth on the surface. Fruits and vegetables which deviate from these general requirements are listed individually in the table.

BIBLIOGRAPHY

ANON. 1948. Interaction of water and porous materials. Discussions Faraday Soc. No. 3.

ANON. 1953. Alcoa aluminum foil—its properties and uses. Aluminum Company of America, Pittsburgh, Penna.

BALL, C. O. 1963. Flexible packaging in food processing. In Food Processing Operations, Vol. II. Edited by M. A. Joslyn and J. L. Heid. Avi Publishing Co., Westport, Conn.

BALL, C. O., CLAUSS, W. E., and STIER, E. F. 1957. Factors affecting quality of prepackaged meats. B. Loss of weight and study of texture. Food Technol. 11, 281–283.

BRICKMAN, C. L. 1957. Evaluating the packaging requirements of a product. Package Eng. 2, No. 7, 19.

BRICKMAN, C. L. 1961. Determining WVTR by new pouch method. Part I. Development of a method. Package Eng. 6, No. 12, 47–51.

BURKINSHAW, L. D. 1964. Polycarbonate film. Modern Packaging 38, No. 3A, 141.

CLARK, F. 1964. Fluorohalocarbon film. Nylon. Modern Packaging 38, No. 3A, 139–140.

CRANE, D. R. 1964. Aluminum foil for packaging. Modern Packaging 38, No. 3A, 163–166.

DAVIS, R. E. 1964. Packaging ideas. Baking Industry 122, No. 1532, 76–78.

DIXON, M. C. 1964. Polythene films for shrink wrapping. Food Trade Rev. 34, No. 2, 48.

DULMAGE, F. C. 1964. Saran. Modern Packaging 38, No. 3A, 143–144.

FELL, M. M. 1964. The Cryovac range of shrink films. Food Trade Rev. 34, No. 2, 48–49.

FLOOD, J. E. 1964. Polyethylene. Modern Packaging 38, No. 3A, 137–138.

HARRIS, A. P. 1964. PVC. Food Trade Rev. 34, No. 2, 47–48.

KIRBY, T. E. 1964. Vinyl films. Modern Packaging 38, No. 3A, 144, 149.

LOWRY, R. D. 1964. The growth of shrink packaging. Modern Packaging 38, No. 3A, 243–250.

LUH, B. S. and TSIANG, J. M. 1965. Packaging of tomato ketchup in plastic laminate and aluminum foil pouches. Food Technol. 19, 395–399.

NELSON, T. J. 1949. Hygroscopicity of sugar and other factors affecting retention of quality. Food Technol. 3, 347–351.

PETERSON, M. S. 1958. Looking at physical principles that explain packaging tests. Package Eng. 3, No. 4, 36–51.

RANKEN, M. D. 1964. Shrinkable packaging of poultry. Food Trade Rev. 34, No. 2, 51–52.

REYNOLDS, C. M. 1964. Cellulose acetate. Modern Packaging 38, No. 3A, 138–139.

SALWIN, H. 1959. Defining minimum moisture contents for dehydrated foods. Food Technol. 13, 594–595.

SALWIN, H. 1959. Moisture levels required for stability in dehydrated foods. Food Technol. 17, 1114–1115, 1118–1120, 1124.

SALWIN, H., and SLAWSON, V. 1959. Moisture transfer in combinations of dehydrated foods. Food Technol. 13, 715–718.

SCHULZ, G. L. 1964. Polystyrene. Modern Packaging 38, No. 3A, 143.

SCOTT, R. 1964. Application of shrink wrapping materials to cheese. Food Trade Rev. 34, No. 2, 52.

SEILER, D. A. L., and KNIGHT, R. A. 1963. Some factors influencing the use of mould inhibitors in cake. Biscuit Maker and Plant Baker 14, No. 7, 600, 602, 604–606, 609–610.

SHERWOOD, W. 1964. Permeability in packaging. Biscuit Maker and Plant Baker 15, No. 8, 624, 628.

SINEATH, H. H. 1964. Cellophane. Modern Packaging 38, No. 3A, 155–158.

STONE, M. C., and REINHART, F. W. 1954. Properties of plastic films. Modern Plastics 31, No. 10, 203–208.

SWIFT, G. 1964. Polypropylene. Food Trade Rev. 34, No. 2, 50–51.

THOMPSON, A. J. 1964. Polyester. Modern Packaging 38, No. 3A, 141–142.

TIERNAN, J. J. 1964. Pliofilm. Modern Packaging 38, No. 3A, 140–141.

WECKEL, K. G., and McCOY, E. 1960. Some causes and effects of moisture translocation in canned baked goods. Report No. 5, 15 July 1960 (Contract DA19-129-OM-1362) Project No. 7-84-06-032. Quartermaster Food and Container Institute for the Armed Forces.

WHITTIER, E. O., and GOULD, S. P. 1930. Vapor pressures of saturated equilibrated solutions of lactose, sucrose, glucose, and galactose. Ind. Eng. Chem. 22, 77–78.

WINTERS, R. M. 1964. Polypropylene. Modern Packaging 38, No. 3A, 142–143.

WRIGHT, R. C., ROSE, D. H., and WHITEMAN, T. M. 1954. The Commercial Storage of Fruits, Vegetables, Ornamentals, and Nursery Stock. U. S. Dept. Agr. Handbook No. 66.

ZENLEA, B. J. 1934. There is a cocoa for every use. Food Inds. 6, 402–403, 434.

CHAPTER 18

The Relation of Microbial Spoilage
to Water Activity of Foods

INTRODUCTION

Microorganisms require an aqueous environment in which to carry on the solute exchanges accompanying growth and reproduction. A microorganism which is not in contact with an external aqueous system of reasonable extent is effectively isolated from its environment even though some gases from endogenous processes may continue to be given off for a time and liquid or solid wastes or products of autolysis may accumulate around the cell. These reactions are seldom responsible for any type of spoilage and may safely be ignored in the present discussion.

On the other hand, it is a matter of common observation that fluid aqueous solutions containing only innocuous components resist spoilage if they are sufficiently concentrated. Evidently the presence of a continuous liquid medium containing a substantial proportion of water molecules is not sufficient in itself to allow detectable growth of microorganisms. The controlling effect of water on spoilage of foods is made use of in preservation by drying, freezing, candying, brining, etc.

The necessary interactions with the environment can occur in media which appear to be solid, as in agar gels and many foodstuffs, but it is essential that there exist some continuous network of water molecules having a form permitting the solution and diffusion of metabolites. It is evident that immobilization of the water molecules to a degree which will prevent such interactions will occur at different moisture contents in different substances. Unrefined cane sugar of two per cent moisture content is known to support microbial spoilage reactions (in the syrup film on the surface of granules) while starch having a considerably higher moisture content is resistant.

Moisture content thus seems to be an inexact indication of the susceptibility of a product to microbial spoilage. A factor which appears to be more closely related to conditions leading to the onset of microbial growth is the water activity of the system. As stated in a previous chapter, the water activity, a_w, of an aqueous system is defined as the ratio of its vapor pressure to that of pure water under identical conditions, $a_w = P_x'/P_w$. Some authors consider measurements based on the amount of water present as vapor (e.g., water activity) to be imprecise as indicators of the availability of water in the cell's immediate environment. Beers (1958)

249

employed the term "moisture activity" to express a quantity representing the capacity of the water present to act as a liquid. Most investigators continue to use equilibrium relative humidities or water activities, and sometimes moisture contents, in reporting data from studies of this type.

Each species of microorganism seems to have a typical minimum water activity at which it can grow. Some molds are the least demanding in this respect while bacteria and yeasts generally require environments of considerably higher water activity before they can initiate growth and reproduction. Latent periods of more than two years before mold spore germination have been observed at low controlled relative humidities (Snow 1949). Generally, increases in water activity above the minimum will permit increased growth, but there are some organisms which will grow as well or better at some higher osmotic value or ion concentration than they will in solutions of higher water activity. These organisms have been called, in different cases, halophilic, osmophilic, xerophytic, etc., (these terms are not interchangeable). In extreme examples, such organisms will not grow at all at water activities near unity.

The limiting water activities of liquid media appear to be at least roughly comparable to the limiting water activities in solid substrates. That is, a mold which would grow on a solution of, say, 0.75 a_w but not on one of 0.70 a_w might be expected to be totally inhibited on a sample of grain having a water activity of 0.70 while it might grow if the sample was raised to a water activity of 0.75. Where significant differences between requirements on liquid and solid media have been found, it has usually been observed that a slightly higher water activity is required for the initiation of growth in solid media than in liquids. This phenomenon is generally attributed to presence of discontinuities in the solids, or other factors not directly related to the measured water activity.

There may be a succession of organisms involved in the spoilage of a product, with the most xerophytic appearing first and, after they have altered the water activity of the substrate by adding moisture generated in oxidative reactions, they are succeeded by more rapidly growing yeasts, bacteria, or molds which flourish under the changed conditions. For example, it has been shown that *Bacillus mesentericus* ("rope" organism) can grow in canned bread at moisture contents inhibitory to *Clostridium botulinum*. After rather extensive alteration of the bread by the former organism has occurred, the water activity of the bread may be raised to a level which would permit the germination and reproduction of Clostridia.

The relationship of moisture content or water activity to the spoilage of cereal grains and their products has probably been studied in greater detail than for other foods. Important economic considerations are involved in this problem since vast quantities of grains are stored at ambient conditions for long periods, their preservation being dependent principally on the maintenance of low moisture contents. Most authorities agree that cereal grains and most products derived from them are stable indefinitely below about 13% moisture content. Above this level, the point at which significant invasion by fungi occurs is dependent upon the type of grain, the temperature, and other factors—or, more fundamentally, upon the water activity. The equilibrium relative humidities of these materials are in the neighborhood of 75% at moisture contents of 15 to 16% as usually determined. Webb et al. (1960) determined the relationship of interparticle relative humidity in feed ingredients and feed mixtures to the growth of molds and spontaneous heating. When the relative humidity was measured at the same temperature as the storage temperature, mold growth and heating did not occur at 72.0% or less during a six-week storage period at 50°, 70°, or 90°F. Snow et al. (1944) determined the safe moisture levels for many feedstuffs, as shown in Table 32. An exemplar of the

TABLE 32

MOISTURE LEVELS BELOW WHICH MOLD SPOILAGE WILL NOT OCCUR IN FEEDSTUFFS[1]

	Per Cent Moisture	
Feed Product	For Short Period Storage	For Long Period Storage
Cereals and Cereal By-products		
Wheat	15.7	14.6
Corn	14.8	13.7
Barley	14.8	13.6
Oats	14.5	13.4
Middlings	14.4	13.1
Bran	14.4	12.8
Legumes		
Locust beans	15.1	11.3
Peas	14.7	13.3
China beans	13.9	12.7
Scotch beans	13.9	12.4
Morocco beans	13.5	12.0
Soybeans	13.3	11.4
Oil Cakes		
Coconut cake	15.1	12.6
Linseed cake	12.3	11.0
Undecorticated cotton cake	12.3	11.0
Undecorticated peanut cake	12.3	11.0
Palm kernel cake	11.5	10.2
Decorticated peanut cake	11.5	9.8
Miscellaneous Feeds		
Malt culms	15.3	12.6
Straw	14.8	12.8
Blood fibrinogen	14.1	12.9
Artificially dried grass	13.7	11.1
Hay	12.6	11.0
Fish meal	11.5	9.9
Distiller's grains	11.0	9.8
Meat and bone meal	10.3	8.7
Bone meal	9.5	8.4

[1] Snow *et al.* (1944).

grain studies which have been conducted in recent years is the very extensive work of Christensen and his co-workers at the University of Minnesota. In a typical paper, Tuite and Christensen (1957) described an experiment in which various samples of wheat nearly or entirely free of fungi were inoculated with four species of the *Aspergillus glaucus* group known to be prevalent on commercially stored grain. Samples were brought to various moisture contents between 12.2 and 16.0% and stored at 77°F. for 1 to 15 months. At moisture contents between 13.0 and 13.6%, the seeds were invaded gradually by *A. restrictus* and *A. amstelodami* while *A. repens* and *A. ruber* acted more slowly. At moisture contents of 14.3 and 14.6%, *A. repens* invaded a larger percentage of seeds in two months than did the other species, but within four months all fungi had invaded all of the seeds held at these moisture levels. Complete invasion occurred within one month at 15.5 to 16.0% moisture. Tuite and Christensen concluded that the upper limit of moisture content for long time storage of wheat at temperatures permitting invasion by storage fungi would be approximately 13%. Similarly, it was found that inoculation of high quality yellow dent corn with mixtures and individual species of storage fungi isolated from corn led to the invasion of the kernels with subsequent killing and discoloration of the germs when the moisture content was above 14% and the temperature above 50°F. Low temperature was as effective as

TABLE 33

LATENT PERIOD OF SPORE GERMINATION (DAYS) OF DIFFERENT MOLDS AT GIVEN HUMIDITIES AT 77°F. ON NUTRIENT GELATIN[1]

Organism	Per Cent Relative Humidity												
	100	95	93	90	88	86	84	82	80	78	76	75	73
Mucor spinosus	1	1	2	..	..	..	..	..	..	..	..	..	..
Mucor sp.	1	1	2	..	..	..	..	..	..	..	..	..	..
Rhizopus nigricans	1	1	2	..	..	..	..	..	..	..	..	..	..
Botrytis cinerea	1	2	2	..	..	..	..	..	..	..	..	..	..
Tricothecium roseum	1	1	2	4	..	..	..	..					
Cladosporium herbarum	1	1	2	2	7	..	..						
Penicillium rugulosum	..	..	..	3	4	9	..						
P. cyclopium	1	1	2	2	3	4	9	..					
Aspergillus niger	1	1	2	2	3	4	11	..					
Penicillium phaeo-janthinellum	..	..	..	3	3	4	7	15	..				
P. wortmanni	..	..	..	2	2	3	4	20	..				
Aspergillus nidulans	..	..	..	1	2	2	4	18	..				
Penicillium sartoryi	..	..	..	2	2	4	4	12	18	..			
P. fellutanum	..	..	..	..	2	4	4	9	12	..			
Aspergillus versicolor	..	..	..	1	2	2	4	7	9	24	..		
A. sydowi	..	..	..	1	2	2	4	7	12	34	..	..	
A. chevalieri var. intermedius	..	..	..	..	..	..	4	7	12	24	53	..	
A. restrictus	..	..	..	..	..	..	3	4	7	12	15	..	
A. candidus	..	..	..	..	..	..	4	7	9	12	15	..	
A. amstelodami	..	..	..	..	..	..	4	4	9	12	18	..	

[1] Snow (1949).

low moisture content (up to 18%, the limit of the study) in preventing damage by the fungi.

The transformation from a totally resistant to a fairly susceptible condition takes place over a narrow moisture range. Within a range of 13 to 15%, an increase of less than one per cent may greatly increase the growth of *Aspergillus restrictus* (Christensen and Linko 1963). An increase of 0.5% in moisture content in the range of 14.2 to 15.5% greatly increased the growth of this mold during an observation period of 36 to 68 days, and of *A. repens* within 170 days. Increasing the moisture content 0.2 or 0.3% in the range from 14.2 to 15.0% resulted in a considerable increase in the growth of *A. restrictus* within 76 to 153 days and of *A. repens* within 153 days.

Mold invasion proceeds very rapidly at moisture contents which permit the proliferation of bacteria and yeast and consequently the latter are of little practical importance in causing grain damage. Fungal attack on cereals begins prior to harvest and probably before the grain has field-dried. In one study by Christensen and Linko (1963), 86 to 100% of the surface-disinfected kernels from many samples of hard red winter wheat were infected with field fungi. Snow (1949) observed the germination of mold spores placed on gelatin plaques held at constant relative humidities. Some of the results are shown in Table 33. No mold spore from the species studied by Snow germinated below 75% R. H.

The optimum water activity for the growth of most bacteria is in the range of 0.995 to 0.990 (Evans and Niven 1960). Their growth rate decreases below this range, but the minimum a_w permitting growth varies

TABLE 34

MINIMUM WATER ACTIVITIES FOR GROWTH OF CERTAIN BACTERIA[1]

Species[2]	a_w
Bacterium mycoides	0.990–0.970
Bacterium pyocyaneum	0.985–0.945
Bacterium asterosporus	0.985
Bacterium luteus	0.985
Bacterium radicicola	0.980–0.965
Azotomonas insolita	0.970
Pseudomonas tumefaciens	0.960
Bacterium mesentericum	0.955
Bacterium vulgare	0.960–0.940
Bacterium coli	0.960–0.935
Bacterium subtilis	0.950
Bacterium prodigiosum	0.945
Bacterium aerogenes	0.945
Mycobacterium siliacum	0.940
Pseudomonas iniqua	0.940
Sarcina sp.	0.930–0.915
Micrococcus roseus	0.905

[1] Burcik (1950).
[2] Original nomenclature retained. From one to five strains of each species were investigated.

widely between different species (see Table 34). For example, the minimum a_w for Salmonellae has been reported near 0.94 while the minimum
for staphylococci is near 0.86 (Evans and Niven 1960). The principal
genera of low temperature bacteria growing on the surfaces of foods, i.e.,
Pseudomonas, Achromobacter, Flavobacterium, and Micrococcus require
relative humidity values approaching saturation (Ayres 1962).

Several yeasts can grow, at least slowly, at relative humidity levels of 85
to 92%. Yeasts of this type are frequently found in the films which develop on the surface of the liquid in pickling vats. They can cause a form
of spoilage of the product.

In their review of yeast taxonomy, Mrak and Phaff (1948) describe film
yeasts isolated from pickling brines of 4 to 20% concentration. *Debar-
yomyces* species had the greatest salt tolerance and were the most common organisms present. *D. membranaefaciens* and its variety *hollandicus*,
and *D. guilliermondii* var. *nova zeelandicus* could be induced to grow in
24% salt brine, corresponding to a water activity of about 0.85. Also
found were *Pichia membranaefaciens* and *Mycoderma decolorans*, the latter having a lower salt tolerance was barely able to grow in 15% salt solution. See Table 35 for the relationship between water activity and the concentration of sodium chloride solutions. Von Schelhorn (1950) states that
a variety of osmophilic yeasts are able to grow at 62% relative humidity.

Scott (1953) clearly demonstrated that the water-related factor controlling growth of *Staphylococcus aureus* in various media at 86°F. was the
water activity and not the moisture content. By varying the concentration
of different solutes in the media, he found minimum water contents (as
per cent dry weight) for growth of 16 to 375%. When the a_w for these
media were determined, they fell in a range of 0.86 to 0.88.

For a given organism, the a_w at which the greatest rate of growth is observed is remarkably constant and often seems to be independent of the

TABLE 35

RELATIONSHIP BETWEEN THE WATER ACTIVITY AND THE CONCENTRATION OF SALT SOLUTIONS[1]

	Sodium Chloride Concentration	
Water Activity	Molal	Per Cent, w/v
0.995	0.15	0.9
0.99	0.30	1.7
0.98	0.61	3.5
0.96	1.20	7
0.94	1.77	10
0.92	2.31	13
0.90	2.83	16
0.88	3.33	19
0.86	3.81	22

[1] Evans and Niven (1960).

solutes present (Marshall and Scott 1958). *Aspergillus amstelodami* grew most rapidly at 0.96 a_w whether the controlling solute was sucrose, glucose, glycerol, sodium chloride, or magnesium chloride (Scott 1957). With the halophilic organism *Vibrio costicolus,* it has also been shown that sodium chloride can be replaced by several other electrolytes at an equivalent "particle concentration," this being the sum, for each of the solutes present, of the product of the molarity and the number of ions per molecule. However, non-electrolytes did not support growth when they were used to adjust the osmotic pressure. Flannery (1956) concluded from the preceding observations that there were two requirements for *V. costicolus*, one osmotic and the other ionic in nature.

Most osmophilic yeasts seem to be rather tolerant to substitution of one solute for another (Ingram 1957). Caution is necessary in extrapolating these results to other organisms since different species may show marked preferences for certain ions which cannot be completely replaced by others. The obligate halophilic bacteria have a specific requirement for high concentrations of sodium ion, according to most authorities. They grow best at sodium chloride concentrations of 20 to 25%, corresponding to water activities of 0.80 to 0.87. Brown and Gibbons (1955) showed that *Halobacterium salinarium* required sodium chloride of at least three molar concentration for growth and that not more than half of this salt could be replaced by potassium chloride. On the other hand, Takahashi and Gibbons (1959) reported that small amounts of calcium or magnesium ions reduced the minimum sodium chloride requirements of *Micrococcus halodenitrificans* from 0.55 to 0.30 M. Generalizations are clearly risky when dealing with osmophilic organisms.

Halophilic bacteria appear to have very concentrated intracellular aqueous phases (Christian and Waltho 1962) and there is evidence that some if not all of their enzymes require the presence of high concentrations of ions for maximum activity. In the extremely halophilic red halophiles studied by Christian and Waltho, the major solutes appear to be sodium chloride and potassium chloride, with the sodium concentration always being below that of the growth medium while the potassium concentration was always much higher than in the medium. In other bacterial groups, sodium and potassium ions probably occur largely in association with amino acids and organic anions.

In predicting the susceptibility of a foodstuff to microbial invasion, the average moisture content, or even the water activity, may not be the critical datum. Although most food raw materials of reasonably stable composition may be expected to have equilibrated throughout fairly well in their natural state at any given time, treatment of any kind, including merely cutting into the surface or any degree of spoilage, may alter the moisture vapor relationships of the different parts of the piece so that equilibrium does not exist.

Fresh fruits and vegetables, as normally collected, will exhibit water activities from 0.98 to almost 1.00. Meat freshly cut will have water activities of about 0.99. Many mixtures and prepared foods (most of which are cooked with added water) will also have equilibrium relative humidities approaching saturation. Since the atmospheres in contact with the surfaces of these products rarely have water activities this high, there is a tendency for water vapor to pass from the object to the atmosphere. This translocation of moisture is restricted by the outer tissues of fruits and vegetables (which may be almost completely enclosed by a waxy cuticle), by relatively impermeable membranes and fatty tissues in meats, and by other structures which hinder the free passage of water vapor. Unless at least one dimension of the object is very small, translocation of vapor from the interior to the dehydrating surface layers can be expected to show a constantly declining rate as the water is withdrawn, even though the decrease may not be measurable by any known technique.

Once invasion is well established, it is difficult to reverse the process by mild drying or by attempts at equilibration since the oxidative reactions of the microorganisms generate water and this water is created precisely at the point where it is most effective, i.e., in the midst of the most actively metabolizing colonies of the bacteria, yeast, or mold.

The situation with respect to fresh meat is somewhat different, according to some authors. The water activity of these products is generally near 0.99, which is about the optimum for many species of bacteria. However, contamination with bacteria is usually restricted to the surface of the meat and drying of the surface will progressively inhibit the growth of most of the spoilage organisms, since the minimum a_w of the most prevalent species is as high as 0.98. The effect of drying on the water activity at the surface of a cut of meat and the consequent effect on microbial spoilage is difficult to express quantitatively. As a product dries, there is a continual decrease in the number of bacterial species that can grow. Also, as the surface dries, there is a slow migration of water from the deeper tissues to the surface and a movement of solutes from the surface to deeper tissues.

Molds are not as susceptible to inhibition by this mild dehydration. Some molds will grow at a_w values as low as 0.75 and meat may have to be dried to a water activity of 0.70 to 0.65 before it can be stored for a long period without mold growth. However, bacterial spoilage ordinarily may be prevented or at least greatly delayed at a water activity of 0.85.

Prevention of spoilage by control of water activities does not necessarily require drying to an unpalatable extent. Although some organisms are capable of growing at water activities that are difficult to achieve without causing undesirable texture changes in the food, these osmophiles can usually be controlled by mild treatments which do not much affect the organoleptic properties of the product. For example, canned bread cannot be processed at temperatures high enough to kill spores of *Clostridium botulinum* because rigorous heat treatment of this food causes texture deterioration, browning, development of off-flavors, and other unacceptable changes. Complete safety of canned bread can be assured by bringing the water activity to levels low enough to eliminate spore-germination and growth of this food poisoning organism. In practice, the Armed Forces maintain control by using a standard formula (by which the concentrations and kinds of solutes in the aqueous phase are established) and by limiting the moisture content of the finished product to not more than 35.0% rather than by measuring the water activity of the bread. Many studies, such as those of Kadavy and Dack (1951) and Wagenaar and Dack (1954) have proved the reliability of this procedure for insuring safety in canned bakery products. The value of this practice has been confirmed by the results—distribution of many millions of cans of bread and similar bakery foods to the Armed Forces during the past 20 years without the occurrence of a single case of food poisoning. Additional insurance against botulism as well as the development of ropiness, is obtained by adjusting the pH to 4.5. Bread prepared according to these specifications is not unpalatable although it is slightly firmer than commercial bread of higher moisture content.

Response of microorganisms to environmental factors such as hydrogen ion concentration and temperature are affected by changes in water activity, and vice versa, at least near the minimum. There appears to be a consensus that the greatest tolerance to low a_w is shown at the optimum pH and temperature for the organism in question. These interactions have been studied by several investigators, among them Wodzinski and Frazier (1960, 1961A, 1961B, 1961C, and 1961D), who showed that cooling of *Pseudomonas fluorescens* or *Aerobacter aerogenes* from 86° to 59°F. resulted in an increase in the minimum a_w under all conditions of testing. Changes at pH 5.4 and 8.8 were greater than at 7.0. Lag periods exhibited similar trends. The growth rate of *Lactobacillus viridescens* is more sensitive to reduced a_w at 99°F. than at 68° or 77°F. These workers also found that *L. viridescens* frequently showed better growth at 0.975 than at 0.992, particularly when either the temperature or the pH was unfavorable. Perhaps the degenerative reactions which might have been promoted by the unfavorable conditions were reduced in rate to a greater extent by the drop in a_w than were the normal processes.

The interrelationship of pH and water activity is of practical importance. As Tressler and Joslyn (1961) have pointed out, honey at a pH of 4 to 5 may ferment even though it contains more than 75% solids, orange concentrates of pH 3.0 to 3.5 do not ferment above about 72% solids, and concentrated lemon juice having a pH near 2.5 seldom ferments at 60% solids.

The maintenance of microorganisms at low relative humidities apparently increases their sensitivity to some destructive agents. Webb (1960) found that relative humidity plays an important part in altering the extent of radiation damage to certain bacterial cells and viruses. As the result of a lengthy series of studies, he concluded that the death of a microbial cell at low water activities might result from a physical change in the structure of an essential macromolecule when the water bound to it is removed. Certain chemicals capable of forming bonds of the proper type (inositol being the most effective) can apparently replace water molecules removed from DNA by drying. In this connection, Webb pointed out that inositol will preserve the activity of dried Rous sarcoma virus, an RNA-virus which exhibits a sharp increase in sensitivity to ultraviolet irradiation when the relative humidity falls below 65%. The special effectiveness of inositol in protecting microorganisms against the destructive action of drying may be the result of its symmetry and structural similarity to water. When inositol was used in the experiments by Webb (1961A, 1961B, and 1961C), the principal function which seemed to persist was the ability to synthesize protein.

Cormack and Morrison (1964) confirmed and extended the work of Webb by studies on *Escherichia coli* and *Pseudomonas aeroginosa* irradiated with ultraviolet of 2,800 to 3,200 Å. and *Serratia marcescens* irradiated with 250 kv. X-rays. Organisms maintained at more than 70% relative humidity suffered very little irradiation or drying damage under the conditions used by Cormack and Morrison. At about 65% there was a rapid change in sensitivity to both ultraviolet light and drying alone. Organisms which had been washed in a five per cent solution of inositol underwent much less inactivation, both in the dark and under ultraviolet irradiation. They suggested that these findings might be due to the beginning of water removal from macromolecules such as DNA at 70% relative humidity. However, *S. marcescens* cells appeared to be stabilized to X-rays by low relative humidity.

The reasons for variations in the minimum water activity for growth of microorganisms is not known with certainty, but Christian and Waltho (1961A), in an investigation of the minimum a_w requirements for growth of 32 non-halophilic bacteria, found a large negative correlation between this factor and the ability to accumulate potassium ions from a standard medium. Christian (1956) showed that halophilic bacteria accumulate high concentrations of potassium and many of the enzymes in such bacteria require high concentrations of salts for maximum activity.

The weight of the evidence is against the hypothesis that the inhibitory action of low a_w is due entirely to the osmotic effect on the cell, since it has been shown that bacteria can grow in media which actually cause dehydration of the organism, as compared to those grown in contact with solutions of high a_w. For *Staphylococcus aureus,* Christian and Waltho (1961B) found that a reduction of the a_w, at which the cells were grown, led to a considerable decrease in the water content per cell, and that similar dehydration was achieved by transferring cells grown in dilute media to more concentrated solutions. These results indicate that bacteria, at least, do not have complete control over the internal water activity unless they are grown in relatively dilute media.

BIBLIOGRAPHY

AYRES, J. C. 1962. Temperature and moisture requirements. *In* Proc. Low Temperature Microbiology Symposium—1961. Campbell Soup Co., Camden, N. J.

BEERS, R. J. 1958. Spores. Vol. II. Edited by H. O. Halvorson. American Institute of Biological Sciences, Washington.

BROWN, H. J., and GIBBONS, N. E. 1955. The effect of magnesium, potassium, and iron on the growth and morphology of red halophilic bacteria. Can. J. Microbiol. *1*, 486–494.

BURCIK, E. 1950. The relationship between hydration and growth in bacteria and yeast. Arch. Mikrobiol. *15*, 203–235.

CHRISTENSEN, C. M. 1963. Influence of small differences in moisture content upon the invasion of hard red winter wheat by *Aspergillus restrictus* and A. *repens.* Cereal Chem. *40*, 385–390.

CHRISTENSEN, C. M., and LINKO, P. 1963. Moisture content of hard red winter wheat as determined by meters and by oven drying, and influence of small differences in moisture content upon subsequent deterioration of the grain in storage. Cereal Chem. *40*, 129–137.

CHRISTIAN, J. H. B. 1956. The physiological basis of salt tolerance in halophilic bacteria. Thesis. Cambridge University.

CHRISTIAN, J. H. B., and SCOTT, W. J. 1953. Water relations of Salmonellae at 30°C. Australian J. Biol. Sci. *6*, 565–575.

CHRISTIAN, J. H. B., and WALTHO, J. A. 1961. The sodium and potassium content of non-halophilic bacteria in relation to salt tolerance. J. Gen. Microbiol. *25*, 97–102.

CHRISTIAN, J. H. B., and WALTHO, J. A. 1962. Solute concentrations within cells of halophilic and non-halophilic bacteria. Biochim. Biophys. Acta *65*, 506–508.

CORMACK, D. V., and MORRISON, H. G. 1964. Relative humidity, inositol, and the effect of radiations on air-dried organisms. Nature *201*, 1103–1105.

EVANS, J. B., and NIVEN, C. F., JR. 1960. Microbiology of meat: Bacteriology. *In* The Science of Meat and Meat Products. W. H. Freeman and Co., San Francisco.

FLANNERY, W. L. 1956. Current status of knowledge of halophilic bacteria. Bacteriol. Rev. 20, 49–66.

GANE, R. 1941. The water content of dried foodstuffs as a function of humidity and temperature. Low Temp. Res. Sta. Interim Record Memo. No. 111.

GINZBURG, B. Z., and Cohen, D. 1964. Calculation of internal hydrostatic pressure in gels from the distribution coefficients of non-electrolytes between gels and solutions. Trans. Faraday Soc. 60, Part 1, 185–189.

GORBANEV, A. U., KESSLER, Y. M., POVAROV, Y. M., SEVASTYANOV, E. S., and KRYLOV, V. V. 1960. The question of "bound" and free" water. Izvest. Akad. Nauk S.S.S.R. Otdel. Khim Nauk 1960, 570–571.

GUILBOT, A. 1949. Considerations on the water content of biological substances: Application to wheat kernels. Compt. rend. acad. agr. France 36, 694–700.

INGRAM, M. 1957. Microorganisms resisting high concentrations of sugars or salts. Symp. Soc. Gen. Microbiol. No. 7, Microbial Ecology, 90–113.

KADAVY, J. L., and DACK, G. M. 1951. The effect of experimentally inoculated canned bread with spores of Clostridium botulinum and Bacillus mesentericus. Food Research. 16, 328–337.

KVAALE, O., and DALHOFF, E. 1963. Determination of the equilibrium relative humidity of foods. Food Technol. 17, 659–661.

LANDROCK, A. H., and PROCTOR, B. E. 1951. A new graphical interpolation method for obtaining humidity equilibrium data, with special reference to its role in food packaging studies. Food Technol. 5, 332–337.

MARSHALL, B. J. and SCOTT, W. J. 1958. The water relations of Vibrio metchnikovi at 30°C. Australian J. Biol. Sci. 11, 171–176.

MRAK, E. M., and PHAFF, H. J. 1948. Yeasts. Ann. Rev. Microbiol. 11, 1–46.

POUNCEY, A. E., and SUMMERS, B. C. L. 1939. The micromeasurement of relative humidity for the control of osmophilic yeasts in confectionery products. J. Soc. Chem. Ind. 58, 162–170.

QASEM, S. A., and CHRISTENSEN, C. M. 1958. Influence of moisture content, temperature, and time on the deterioration of stored corn by fungi. Phytopathology 48, 544–549.

ROBINSON, R. A., and STOKES, R. H. 1955. Electrolyte Solutions. Butterworths, London.

SALWIN, H. 1963. Moisture levels required for stability in dehydrated foods. Food Technol. 17, 1114–1123.

SCHACHINGER, L., and HEISS, R. 1951. Osmotic values and growth of microorganisms in sugar solutions. Arch. Mikrobiol. 16, 347–357.

SCOTT, W. J. 1953. Water relations of Staphylococcus aureus at 30°C. Australian J. Biol. Sci. 6, 549–564.

SCOTT, W. J. 1957. Water relations of food spoilage organisms. Adv. Food Res. 7, 83–127.

SCOTT, W. J. 1962. Available water and microbial growth. In Proc. Low Temperature Microbiology Symposium 1961. Campbell Soup Co., Camden, N. J.

SNOW, D. 1949. The germination of mould spores at controlled humidities. Ann. Appl. Biol. 36, 1–13.

SNOW, D., CRICHTON, M. H. G., and WRIGHT, N. C. 1944. Mould deterioration of feeding-stuffs in relation to humidity of storage. Part II. The water uptake of feeding-stuffs at different humidities. Ann. Appl. Biol. 31, 111–116.

TAKAHASHI, I., and GIBBONS, N. E. 1959. Effect of salt concentration on the morphology and chemical composition of *Microccus halodenitrificans*. Can. J. Microbiol. 5, 25–35.

TUITE, J. F., and CHRISTENSEN, C. M. 1957. Grain storage studies. XXIV. Moisture content of wheat seed in relation to invasion of the seed by species of the *Aspergillus glaucus* group, and effect of invasion upon the germination of the seed. Phytopathology 47, 323–327.

VAISEY, E. B. 1954. Osmophilism of *Sporendonema epizoum*. J. Fisheries Res. Can. 11, 901–903.

VON SCHELHORN, M. 1950. Research into the spoilage of low moisture foods by osmophilic organisms. I. Spoilage by osmophilic yeasts. Z. Lebensm. Untersuch. u. Forsch. 91, 117–124.

WAGENAAR, R. O., and DACK, G. M. 1954. Further studies on the effect of experimentally inoculating canned bread with spores of *Clostridium botulinum*. Food Research 19, 521–529.

WEBB, B. D., BAYLISS, M. E., and RICHARDSON, L. R. 1960. Relation of interspace relative humidity to growth of molds and heating in feed ingredients and feed mixtures. J. Agr. Food Chem. 8, 371–374.

WEBB, S. J. 1959. Factors affecting the viability of air-borne bacteria. I. Can. J. Microbiol. 5, 649–669.

WEBB, S. J. 1960. Factors affecting the viability of air-borne bacteria. III The role of bound water and protein structure in the death of air-borne cells. Can. J. Microbiol. 6, 89–105.

WEBB, S. J. 1961A. Factors affecting the viability of air-borne bacteria. IV. The inactivation and reactivation of air-borne *Serratia marcescens* by ultraviolet and visible light. Can. J. Microbiol. 7, 607–619.

WEBB, S. J. 1961B. Factors affecting the viability of air-borne bacteria. V. The effect of desiccation on some metabolic systems of *Escherichia coli*. Can. J. Microbiol. 7, 621–632.

WEBB, S. J. 1963. Factors affecting the viability of air-borne bacteria. VI. The action of inositol on lactose oxidation by desiccated *Escherichia coli*. Can. J. Biochem. and Physiol. 41, 455–460.

WODZINSKI, R. J., and FRAZIER, W. C. 1960. Moisture requirements of bacteria. I. Influence of temperature and pH on requirements of *Pseudomonas fluorescens*. J. Bacteriol. 79, 572–578.

WODZINSKI, R. J., and FRAZIER, W. C. 1961A. Moisture requirements of bacteria. II. Influence of temperature, pH, and malate concentration requirements of *Aerobacter aerogenes*. J. Bacteriol. 81, 353–358.

WODZINSKI, R. J., and FRAZIER, W. C. 1961B. Moisture requirements of bacteria. III. Influence of temperature, pH, and malate and thiamine concentration on requirements of *Lactobacillus viridescens*. J. Bacteriol. 81, 359–365.

WODZINSKI, R. J., and FRAZIER, W. C. 1961C. Moisture requirements of bacteria. IV. Influence of temperature and increased partial pressure of carbon dioxide on requirements of three species of bacteria. J. Bacteriol. 81, 401–408.

WODZINSKI, R. J., and FRAZIER, W. C. 1961D. Moisture requirements of bacteria. V. Influence of temperature and decreased partial pressure of oxygen on requirements of three species of bacteria. J. Bacteriol. 81, 409–415.

Glossary

A compilation from many sources of words and phrases commonly encountered in works on water engineering and allied fields.

Acidulous water—water which contains relatively large quantities of carbon dioxide or sulfates so that it gives an acidic reaction.

Acre-foot—the quantity of water or sewage required to cover one acre to a foot in depth, or 43,560 sq. ft.

Adjutage—a tube inserted into an orifice.

Aerator—a device to increase the interfacial surface between water and air or sewage and air.

Aerochlorination—the application of compressed air and chlorine gas to water or sewage to accomplish simultaneously chlorination and the removal of grease.

Aftergrowth—growth of microorganisms in a water supply or system following treatment.

Afterprecipitation—the precipitation of colloidal calcium carbonate in a sand filter or distribution system after the water has been treated with lime.

Air wash—a process employing compressed air as an adjunct to water in the washing of the filtering medium of a rapid sand filter.

Alkalinity—as this term is used in water treatment technology, it refers to the content of carbonates, bicarbonates, hydroxides, and occasionally also to borates, silicates, and phosphates. It is expressed as p.p.m. calcium carbonate.

Antibiosis—biological action which inhibits putrefaction and oxidation.

Antichlors—compounds and ions such as sulfur dioxide and bisulfite which can convert excess chlorine into an inert salt.

Aquifer—an underground stratum saturated with water.

Aquifuge—a rock formation which can neither absorb nor transmit water because it does not contain interconnected voids.

Artesian—a ground water formation under pressure sufficient to cause the water to rise to a higher level when an outlet is offered.

***B. coli* index**—in testing for the coliform group of bacteria, this is the number obtained by taking the reciprocal of the highest dilution in a decimal series which gave a positive result. It is commonly referred to as the Phelps index.

Backsiphonage—a flow of polluted or contaminated water into a supply line from a cross-connection with a line containing sewage or untreated water.

Backwashing—the method of cleaning a rapid or mechanical filter by reversing the flow of water through it.

Basin—a natural or man-made structure for holding water.

Benthal demand—the consumption of dissolved oxygen due to the upward diffusion of decomposition products from benthal deposits into higher water levels.

Benthal deposits—the organic materials which accumulate on the beds of streams and other watercourses.

Biochemical oxygen demand (B.O.D.)—the quantity of oxygen utilized in the oxidation of the organic matter in a water sample at a specific temperature and in a specified length of time.

Biological filtration—a method of water purification employing filter bed media encapsulated (at least in part) with zooglealfilms which retain smaller particles than would otherwise be held.

Carbonator—a device for introducing carbon dioxide into a water supply.

Cavitation—the formation of pockets of vapor and gas in a body of liquid. In non-boiling fluids, it is caused by local lowering of pressure in turbulent flow and occurs principally at the axes of vortices.

Chlorinated copperas—ferrous sulfate which has been treated with chlorine gas. It contains ferric chloride as well as ferrous sulfate and, probably, other compounds.

Chlorinated lime—a combination of slaked lime and chlorine gas. It is also called bleaching powder, chloride of lime, and hypochlorite of lime.

Chlorine demand—the difference between the amount of chlorine added to water, industrial waste, or sewage and the amount of residual chlorine remaining at the end of a specified contact period.

Chlorine index—the quantity of chlorine in sewage in relation to the quantity of chlorine in the water supply.

Chlorine residual—the quantity of chlorine in excess of the chlorine demand remaining in the treated material after a selected period of time.

Chlorinity—the total amount of chlorine in grams contained in one kilogram of sea water after the bromide and iodide have been replaced by chloride.

Clarifier—a settling or sedimentation tank.

Coefficient of permeability (standard)—the rate of flow of water, in gallons per minute, through a cross-section one square foot in area, of a porous mass under a unit hydraulic gradient at a temperature of 60°F.

Consecutive digestion—digestion of sludge first under thermophilic conditions and then under mesophilic conditions, followed by concentration.

Contaminated water—a water which is specifically polluted by the introduction into otherwise potable water of deleterious substances such as toxic compounds or bacteria.

Copperas—ferrous sulfate.

Cryology—the science of the physical characteristics of the solid forms of water (such as ice, snow, sleet, etc.) produced by temperatures below 32°F.

Cuprichloramine—an algicide made by combining stoichiometric proportions of copper sulfate, ammonia, and chlorine.

Deferrization—the removal of soluble iron compounds from water.

Denitrification—the reduction of dissolved nitrates by biochemical action.

Detention period—for sedimentation basins, is defined as the volume of the basin divided by the volumetric rate of flow through the basin.

Digestion—as applied to sewage and industrial wastes, is the anaerobic decomposition of organic matter, resulting in partial gasification, liquefaction, and mineralization.

Dispersion—the mixing of polluted fluids with a large volume of water in a stream, lake, etc., or the outward percolation of water from an artesian basin or aquifer through confining formations.

Diversion—the act of taking water from a natural surface source and transferring it into a conduit, pipe, or other transfer system.

Drawdown—the change in the level of water resulting from the withdrawal of supplies from wells or bodies of water.

Drifting sand filter—a rapid sand filter designed in such a manner that the sand comprising the bed drifts from an entrance point to an exit point from which it is drawn off to be washed.

Eductor—a device for mixing air with water.

Effluent—liquid which flows out of a confined space.

Elutriation—a sludge-conditioning process in which various constituents are removed by successive decantations with fresh water or plant effluent, thereby reducing the demand for conditioning chemicals.

Filament—an imaginary threadlike element of a stream of water moving downstream with all particles of the water which comprise it remaining within its limits at all times.

Fluviation—a term applied to the activities of streams.

Free chlorine residual—that portion of the total chlorine residual remaining in the water which reacts chemically and biologically as hypochlorous acid or as hypochlorite.

Geohydrology—that branch of hydrology relating specifically to subsurface or subterranean water.

Geotome—a device for measuring moisture in soils.

Grade—the slope (e.g., of the channel of a stream) usually expressed in terms of the ratio of the rise or the fall to the horizontal distance.

Greensand—glauconite, a natural zeolite.

Grit—the large hard particles in waste effluent—sand, gravel, etc.

Ground water—water occupying the zone of saturation of the soil.

Ground water artery—a mass of permeable material enveloped by strata of less permeable rock and saturated with water that is under pressure.

Hardness—a characteristic of some waters which is related to their interaction with soaps and results from the presence of calcium, iron, and magnesium carbonates and sulfates, with occasional contributions by the nitrates and chlorides of these cations.

High rate filter—a trickling filter operated at the high average throughput of 10 to 30 m.g.d./acre, sometimes with recirculation of the effluent.

Hydraulic elements—the parameters affecting a particular stage of flowing water or a particular cross-section of a conduit or channel; among the more important hydraulic elements would be the depth of the water, velocity, energy head, and friction factor.

Hydraulic gradient—the ratio of the loss in the sum of the pressure head and the position head to the flow distance.

Hydraulic model—a flow system by whose operation the characteristics of a similar (usually larger) system may be predicted.

Hydraulics—that branch of engineering science dealing primarily with the application of fluid mechanics to the flow of water.

Hydrodynamics—the science of fluids in motion.

Hydrograph—a chart showing the stage, velocity, or other function of the discharge of a stream at a given point.

Hydrographic survey—an instrumental survey for determining the characteristics of streams and other bodies of water within an area of interest; it will include such data as the location, depth, and area of water in oceans, the width, depth and course of streams, the positions and elevations of high water marks, the location and depth of wells, etc.

Hydrography—the science of the measurement and analysis of the flow of water, precipitation, evaporation, and related phenomena.

Hydrologic cycle—the complete cycle of natural phenomena involved in the passage of water from the vapor form into precipitation and through the various other stages back into vapor.

Hydrology—the science or classified body of knowledge pertaining to the properties, distribution, and behavior of water in nature; that branch of physical geography dealing with the waters of the earth with special reference to properties, phenomena, and distribution. It treats specifically of the occurrence of water on the earth, the relation of water to life on the earth, the physical effects of the water on the earth, and the description of the earth with respect to water.

Hydrostatic level—the level or elevation to which the top of a column of water would rise from an artesian aquifer or other pressurized source.

Hygrometer—an instrument for determining the dew point.

Impervious—descriptive of material, particularly rock strata, through which water passes very slowly, if at all.

Infected—as applied to water, means contaminated with pathogenic organisms.

Infiltration—the flow or movement of water through the interstices of a soil or other porous medium.

Infiltration gallery—an underground structure designed for the purpose of collecting subsurface water from a relatively wide area.

Influent—the liquid which flows into a confined space such as a reservoir, basin, or treatment plant.

Interstice—a pore or open space in rock, soil, or granular material.

Isobath—a line connecting all points on a map which are the same vertical distance above the upper or lower surface of a water-bearing formation or aquifer.

Isopluvial line—a line connecting all points on a map indicating areas of equal precipitation.

Jetting—use of a water jet to sink holes in the earth for various purposes.

Lake bloom—large masses of plant life (such as algae) which develop on or near the surface of reservoirs, etc.

Laminar flow—flow in which each particle of the liquid moves in a direction parallel to every other particle; also called streamline flow.

Lime and soda ash process—a method for softening water by adding lime and soda ash to form the insoluble compounds, calcium carbonate and magnesium hydroxide.

Limnology—the branch of hydrology specifically concerned with lakes; also, study of the biological productivity of inland waters and the factors affecting it.

Lithosphere—the solid part of the earth.

Marginal chlorination—the addition of small amounts of chlorine without regard to the type of residual produced or the persistence of the residual.

Membrane potential—the difference in potential in a semipermeable membrane system, as measured by two calomel electrodes placed in solutions of different concentrations separated by the membrane.

Meteoritic water—water in, or derived from, the atmosphere; sometimes used to include all subsurface water of external origin and sometimes only water derived by absorption.

Meteorology—the science of the atmosphere; the study of atmospheric phenomena.

Meter, Venturi—a flow-measuring meter for closed conduits, in which the Venturi tube is used in conjunction with a proprietary registering device.

Methyl orange alkalinity—the alkalinity of water measured by the quantity of standardized sulfuric acid solution required to bring a sample of the water to an acid pH, as indicated by the color change of methyl orange; it is expressed as p.p.m. of calcium carbonate.

Moisture tension—a numerical index of the energy with which water is held in a sample of soil; it is expressed on a log scale called pF.

Most probable number (M.P.N.)—in the determination of bacterial numbers by the dilution method, the number of organisms per unit volume which, in accordance with statistical theory, would be more likely than any other number to yield the observed test result or which would yield the observed test result with the greatest frequency. It is expressed as the number of organisms per 100 ml.

Normalization—the removal of carbon dioxide from a water supply to reduce its corrosive action.

Odor threshold—the dilution at which the odor of a water sample can just be detected when tested under standard conditions; specially prepared odorless water is used for diluting the sample.

Phenolphthalein alkalinity—a measurement of the alkalinity of a water performed by titrating with sulfuric acid until the color of the indicator phenolphthalein changes (about pH 8.2). Said to be a measure of the hydroxides plus one-half of the normal carbonates.

Phreatic line—the upper boundary of the water table.

Piezometric surface—an imaginary surface that everywhere coincides with the static level of water in an aquifer or artesian basin.

Pitot tube—a device consisting of an orifice held at a point upstream in flowing water and connected with a tube in which the rise of water due to the velocity head may be measured.

Plain sedimentation—the subsidence and deposition of suspended matter unaided by chemical treatment or other artificial means.

Plate count—the number of visible microbial colonies found on inoculated plates of nutritional media which have been incubated under specified conditions.

Polluted water—a water fouled by materials such as sewage which make it offensive to sight and smell.

Porosity—degree of perviousness of a soil or rock strata; the ratio, usually expressed as a percentage, of (a) the volume of interstices in a given quantity of material to (b) the total volume of the material.

Potable water—water suitable for drinking and culinary purposes.

Primary treatment—the first majoring processing operation applied to sewage in a treatment works—usually sedimentation.

Psychrometry—that body of knowledge and scientific practice dealing with mixtures of air and water vapor.

Purification—the removal by natural or artificial methods, of objectionable matter from water.

Putrescibility—the term used in water or sewage analysis to define stability of a polluted water or partially treated sewage.

Quicklime—a calcined material comprised mostly of calcium oxide or calcium oxide in natural association with a lesser amount of magnesium oxide.

Recarbonation—the bubbling of carbon dioxide through water to restore the carbon dioxide removed when lime was added to soften the water.

Recirculation—the refiltration of all or part of the effluent in a high rate trickling filter for the purpose of maintaining the flow within desired limits or of reducing the strength of the combined effluent.

Reductant—organic matter in streams and sewage which is stabilized under anaerobic conditions.

Reynold's critical velocity—the velocity in a conduit at which the flow of contained liquid changes from the laminar to turbulent; at this point friction ceases to be proportional to the first power of the velocity and becomes proportional to a higher power.

Rippl diagram—a diagram of the net yield of a water supply during a series of consecutive time intervals, used for the study of storage.

Salinity—defined as the total amount of solid material in grams contained in one kilogram of sea water when all of the carbonate has been converted to oxide, the bromine and iodine replaced by chlorine, and all organic matter completely oxidized. The symbol is 0/00.

Sanitary survey—an investigation of any condition affecting the public health but commonly used in a more restricted sense to refer to an examination of the public health aspects of an actual or potential water supply.

Saturation index—a measure of the calcium carbonate in water.

Saturation zone—that portion of the lithosphere in which the functional interstices of permeable rock or earth are filled with water under hydrostatic pressure.

Scott-Darcy process—a precipitation or flocculation technique in which the crude ferric chloride used as a coagulant is made by passing an aqueous solution of chlorine over iron scrap.

Secondary sewage treatment—the treatment of sewage by biological methods after primary treatment by sedimentation.

Sedgwick-Rafter method—a method for the quantitative determination of microscopic organisms larger than bacteria.

Sedimentation—the process of subsidence and deposition resulting from the influence of gravity on suspended matter carried by water, sewage, industrial wastes, and the like.

Sedimentation tanks—tanks or basins for retaining liquids long enough to allow the larger particles to settle to the bottom; also called settling tanks, clarification tanks, and subsidence tanks.

Septicization—an anaerobic decomposition by which solid organic matter, as in sewage, is liquefied and gasified by bacterial action.

Settleable solids—suspended solids which will deposit out of quiescent liquids in a period commonly taken at two hours.

Settling—the precipitation of floc and other material out of coagulated water.

Sewer—a pipe or conduit, normally closed, but usually not flowing full, for carrying sewage and other waste liquids.

Siphon trap—a trap for waste-water lines having a double-bend, like an S, the lower curve containing a water seal which prevents reflux of gases.

Sludge—the semiliquid mass composed of the accumulated settled solids deposited from sewage or industrial wastes which have been held for a time in basins.

Sludge bulking—an undesirable condition in which sludge occupies excessive volumes and will not concentrate readily.

Sludge index—the volume in milliliters occupied by aerated mixed liquor containing one gram of dry solids after settling 30 min.; also called the Mohlman index.

Soda ash—sodium carbonate.

Soil water—water which exists in the zone of aeration immediately below the surface of the ground.

Solifluction—the slow flowing from higher to lower ground of masses of waste saturated with water.

Specific absorption—the capacity of water bearing material to absorb water after the gravity water has been removed.

Specific capacity—as applied to wells, this term describes the rate at which water may be drawn from a formation to cause a drop in level of a specified amount.

Stability—the tendency of a material such as sewage to resist putrefaction.

Stable effluent—treated sewage containing enough oxygen to satisfy its oxygen demand.

Stage—the elevation of a water surface above its minimum level or above some other reference point.

Standard B.O.D.—biochemical oxygen demand as determined under standard laboratory conditions by incubating for five days at 58°F. and usually reported as p.p.m.

Standard coefficient of permeability—the rate of flow of water in gallons per minute through a cross-section of one square foot of porous substance under a unit hydraulic gradient at a temperature of 60°F.

Standard rate filter—a trickling filter having a low rate of throughput per unit of surface area, frequently 1 to 4 m.g.d./acre.

Storage—the impounding of water for considerable periods either in surface or underground reservoirs for future use.

Suspended solids—when used to describe the results of a laboratory determination, this term refers to the quantity of material deposited from a given volume of water or other liquid when it is filtered through an asbestos mat in a Gooch crucible.

Suspensoids—colloidal particles which will remain in suspension through all ordinary processes of water treatment.

Tank—any artificial receptacle through which liquid passes or in which a liquid is held in reserve or detained for any purpose.

Tape gage—a tagged or indexed chain, tape, or other line, attached to a weight which is lowered until the weight touches the water surface.

Temporary hardness—hardness resulting from the presence of dissolved calcium and magnesium carbonates and bicarbonates; called temporary hardness because most of it can be removed by boiling the water.

Tensiometer—an instrument consisting of a porous clay cell filled with water and connected to a manometer; it is used for measuring the potential of soils or for indicating the amount and status of moisture in soils.

Total head—the velocity head and the friction head added to the difference in elevation between the surface of the water at the source of supply and the surface of the water at the outlet.

Transport number—as applied to counterions, the fraction of the current carried by a particular ion.

Tube—in hydraulics, a short pipe attached to an orifice or used to connect two vessels.

Turbidity—in general, a condition of a liquid due to fine particles in suspension causing reflection and refraction of light and decreasing the optical transmittance of the liquid; specifically, a measure of the fine suspended material (usually colloidal) in water.

Turbulence—a state of flow of water marked by cross current and eddy formation as well as other deviations from laminar flow patterns.

Turbulent velocity—the velocity of water flowing in a conduit above which the flow will always be turbulent and below which the flow may either be turbulent or be laminar depending upon other conditions.

Vent—in plumbing, a pipe-line installed to provide a flow of air to or from a drainage system, or to provide a circulation of air within such a system to protect trap seals from siphonage and back pressure.

Water of compaction—water furnished by destruction of pore space owing to compression of sediments.

Water of dilation—water in excess of water of saturation held by sedimentary material in an expanded state.

Water of saturation—water which completely fills the interstices of rock formation or of earth whose particles are in contact (i.e., not expanded).

Watercourse—a natural or artificial channel in which a flow of water occurs in a definite direction either continuously or intermittently; but, if the latter, with some degree of regularity.

Watershed—the area contained within a divide above a specific point on a stream.

Weir—a diversion dam.

Wet-bulb temperature—the dynamic equilibrium temperature attained by a water surface when the rate of heat transfer to the surface by convection equals the rate of mass transfer away from the surface.

Zeolite—a hydrous silicate mineral useful in water engineering because of its ion-exchange properties.

Zooglea—a jelly-like matrix usually associated with activated sludge growths in biological treatment beds.

Index

www.ingramcontent.com/pod-product-compliance
Lightning Source LLC
Chambersburg PA
CBHW031921190326
41519CB00007B/373